DECIPHERING QUANTUM MECHANICS

KEY CONCEPTS EXPLAINED FOR BEGINNERS

DAVID SANDUA

Deciphering Quantum Mechanics.
© David Sandua 2024. All rights reserved.
eBook & Paperback Edition.

"Quantum mechanics is the physics of the possible: reality is a dream."

Albert Einstein

INDEX

I. INTRODUCTION TO QUANTUM MECHANICS 13
- DEFINITION OF QUANTUM MECHANICS 14
- IMPORTANCE IN MODERN PHYSICS 15
- OVERVIEW OF ESSAY STRUCTURE 16

II. HISTORICAL CONTEXT 17
- EARLY 20TH CENTURY PHYSICS 18
- KEY SCIENTISTS AND THEIR CONTRIBUTIONS 19
- MILESTONES IN QUANTUM THEORY DEVELOPMENT 20

III. WAVE-PARTICLE DUALITY 21
- CONCEPT OF DUALITY 22
- EXPERIMENTS DEMONSTRATING DUALITY 23
- IMPLICATIONS FOR UNDERSTANDING MATTER 24

IV. THE QUANTUM STATE AND SUPERPOSITION 25
- EXPLANATION OF QUANTUM STATES 26
- PRINCIPLE OF SUPERPOSITION 27
- THOUGHT EXPERIMENTS AND ILLUSTRATIONS 28

V. QUANTUM ENTANGLEMENT 29
- DEFINITION AND BASIC PRINCIPLES 30
- EPR PARADOX AND BELL'S THEOREM 31
- IMPLICATIONS FOR INFORMATION THEORY 32

VI. THE UNCERTAINTY PRINCIPLE 33
- HEISENBERG'S CONTRIBUTION 34
- MEASUREMENT AND PRECISION LIMITS 35
- PHILOSOPHICAL IMPLICATIONS 36

VII. THE WAVE FUNCTION 37
- ROLE IN QUANTUM MECHANICS 38
- INTERPRETATION OF THE WAVE FUNCTION 39
- PROBABILITY AND MEASUREMENT 40

VIII. THE SCHRÖDINGER EQUATION 41
- INTRODUCTION TO THE EQUATION 42
- TIME-DEPENDENT VS. TIME-INDEPENDENT FORMS 43
- CONCEPTUAL UNDERSTANDING WITHOUT MATH 44

IX. QUANTUM TUNNELING 45
- PHENOMENON EXPLANATION 46
- APPLICATIONS IN ELECTRONICS 47
- CONCEPTUAL CHALLENGES 48

X. QUANTUM COMPUTING 49
- BASICS OF QUANTUM COMPUTERS 50
- QUBITS AND QUANTUM SUPREMACY 51
- POTENTIAL IMPACT ON SOCIETY 52

XI. QUANTUM CRYPTOGRAPHY 53
- PRINCIPLES OF QUANTUM ENCRYPTION 54

QUANTUM KEY DISTRIBUTION (QKD) 55
FUTURE OF SECURE COMMUNICATION 56
XII. QUANTUM TELEPORTATION 57
THEORETICAL FRAMEWORK 58
EXPERIMENTAL REALIZATIONS 59
MISCONCEPTIONS AND CLARIFICATIONS 60
XIII. MEASUREMENT PROBLEM 61
THE OBSERVER EFFECT 62
COLLAPSE OF THE WAVE FUNCTION 63
INTERPRETATIONS AND CONTROVERSIES 64
XIV. COPENHAGEN INTERPRETATION 65
OVERVIEW OF THE INTERPRETATION 66
PHILOSOPHICAL UNDERPINNINGS 67
CRITICISMS AND ALTERNATIVES 68
XV. MANY-WORLDS INTERPRETATION 69
EXPLANATION OF EVERETT'S THEORY 70
IMPLICATIONS FOR REALITY 71
DEBATE AND ACCEPTANCE 72
XVI. DECOHERENCE THEORY 73
ROLE IN QUANTUM MECHANICS 74
EXPLANATION OF CLASSICAL TRANSITION 75
RESOLVING MEASUREMENT PROBLEMS 76
XVII. QUANTUM FIELD THEORY 78
INTRODUCTION TO FIELDS IN QUANTUM MECHANICS 79
UNIFICATION OF FORCES 80
STANDARD MODEL OF PARTICLE PHYSICS 81
XVIII. QUANTUM GRAVITY AND THE SEARCH FOR UNIFICATION 82
CHALLENGES IN UNIFYING GRAVITY WITH QUANTUM MECHANICS 83
APPROACHES TO QUANTUM GRAVITY 84
SIGNIFICANCE FOR COSMOLOGY 85
XIX. PHILOSOPHICAL IMPLICATIONS 86
REALITY AND OBJECTIVITY 87
DETERMINISM AND FREE WILL 88
QUANTUM MECHANICS AND CONSCIOUSNESS 89
XX. QUANTUM MECHANICS IN POPULAR CULTURE 90
MISCONCEPTIONS AND EXAGGERATIONS 91
INFLUENCE ON LITERATURE AND FILM 92
PUBLIC UNDERSTANDING AND INTEREST 93
XXI. EDUCATIONAL APPROACHES TO QUANTUM MECHANICS 94
TEACHING COMPLEX CONCEPTS 95
USE OF SIMULATIONS AND VISUALIZATIONS 96
ENCOURAGING INTUITIVE UNDERSTANDING 97
XXII. QUANTUM MECHANICS AND METAPHYSICS 98
INTERPLAY BETWEEN PHYSICS AND PHILOSOPHY 99
QUESTIONS ABOUT THE NATURE OF EXISTENCE 100
IMPACT ON THEOLOGICAL AND METAPHYSICAL THOUGHT 101
XXIII. QUANTUM MECHANICS IN BIOLOGY 102

QUANTUM EFFECTS IN BIOLOGICAL SYSTEMS 103
QUANTUM BIOLOGY RESEARCH 104
IMPLICATIONS FOR THE STUDY OF LIFE 105

XXIV. QUANTUM MECHANICS AND CHEMISTRY 106
CHEMICAL BONDING AND REACTIONS 107
QUANTUM CHEMISTRY AND COMPUTATIONAL METHODS 108
ADVANCES IN MATERIAL SCIENCE 109

XXV. QUANTUM MECHANICS AND ASTROPHYSICS 110
QUANTUM PHENOMENA IN SPACE 111
BLACK HOLES AND QUANTUM INFORMATION 112
QUANTUM COSMOLOGY 113

XXVI. QUANTUM MECHANICS AND THERMODYNAMICS 114
QUANTUM STATISTICAL MECHANICS 115
ENTROPY AND INFORMATION 116
QUANTUM THERMODYNAMIC PROCESSES 117

XXVII. QUANTUM MECHANICS AND INFORMATION THEORY 118
QUANTUM INFORMATION SCIENCE 119
ENTANGLEMENT AND INFORMATION TRANSFER 120
QUANTUM ALGORITHMS 121

XXVIII. QUANTUM MECHANICS AND MATHEMATICS 122
MATHEMATICAL FOUNDATIONS 123
ROLE OF SYMMETRY AND GROUP THEORY 124
TOPOLOGICAL QUANTUM SYSTEMS 125

XXIX. QUANTUM MECHANICS AND NONLOCALITY 126
CONCEPT OF NONLOCAL INTERACTIONS 127
TESTS OF NONLOCALITY 128
PHILOSOPHICAL AND THEORETICAL IMPLICATIONS 129

XXX. QUANTUM MECHANICS AND DETERMINISM 130
DETERMINISTIC VS. PROBABILISTIC NATURE 131
HIDDEN VARIABLES THEORIES 132
IMPLICATIONS FOR PREDICTABILITY 133

XXXI. QUANTUM MECHANICS AND THE MIND 134
THEORIES OF CONSCIOUSNESS 135
QUANTUM BRAIN DYNAMICS 136
CONTROVERSIES AND SPECULATIONS 137

XXXII. QUANTUM MECHANICS AND ART 138
ARTISTIC INTERPRETATIONS OF QUANTUM CONCEPTS 139
INFLUENCE ON VISUAL AND PERFORMING ARTS 140
DIALOGUES BETWEEN ARTISTS AND PHYSICISTS 141

XXXIII. QUANTUM MECHANICS AND ECONOMICS 142
QUANTUM DECISION THEORY 143
APPLICATIONS IN FINANCIAL MARKETS 144
ECONOMIC MODELING AND QUANTUM SYSTEMS 145

XXXIV. QUANTUM MECHANICS AND ENVIRONMENTAL SCIENCE 146
QUANTUM EFFECTS IN CLIMATE SYSTEMS 147
QUANTUM SENSORS AND ENVIRONMENTAL MONITORING 148
SUSTAINABLE ENERGY TECHNOLOGIES 149

XXXV. QUANTUM MECHANICS AND NANOTECHNOLOGY 150
- NANOSCALE QUANTUM PHENOMENA 151
- QUANTUM DOTS AND NANODEVICES 152
- FUTURE OF NANOSCIENCE 153

XXXVI. QUANTUM MECHANICS AND ENGINEERING 154
- QUANTUM ENGINEERING DISCIPLINES 155
- QUANTUM MATERIALS AND FABRICATION 156
- CHALLENGES IN QUANTUM DEVICE DESIGN 157

XXXVII. QUANTUM MECHANICS AND EDUCATION 158
- CURRICULUM DEVELOPMENT FOR QUANTUM PHYSICS 159
- INNOVATIVE TEACHING METHODS 160
- PREPARING STUDENTS FOR A QUANTUM FUTURE 161

XXXVIII. QUANTUM MECHANICS AND INTELLECTUAL PROPERTY 162
- PATENTING QUANTUM TECHNOLOGIES 163
- LEGAL AND ETHICAL CONSIDERATIONS 164
- IMPACT ON INNOVATION AND RESEARCH 165

XXXIX. QUANTUM MECHANICS AND GLOBAL SECURITY 166
- QUANTUM COMPUTING AND CRYPTOGRAPHY IN DEFENSE 167
- NONPROLIFERATION OF QUANTUM WEAPONS 168
- INTERNATIONAL AGREEMENTS AND REGULATIONS 169

XL. QUANTUM MECHANICS AND SPACE EXPLORATION 170
- QUANTUM SENSORS IN SPACECRAFT 171
- QUANTUM COMMUNICATION IN SPACE 172
- IMPLICATIONS FOR INTERSTELLAR TRAVEL 173

XLI. QUANTUM MECHANICS AND PHILOSOPHY OF SCIENCE 174
- SCIENTIFIC REALISM AND INSTRUMENTALISM 176
- THEORY CHANGE AND SCIENTIFIC REVOLUTIONS 177
- QUANTUM MECHANICS AND SCIENTIFIC METHOD 178

XLII. QUANTUM MECHANICS AND LITERATURE 179
- LITERARY THEMES INSPIRED BY QUANTUM THEORY 180
- SCIENCE FICTION AND QUANTUM MECHANICS 181
- NARRATIVE STRUCTURES AND QUANTUM CONCEPTS 182

XLIII. QUANTUM MECHANICS AND GENDER STUDIES 183
- GENDER PERSPECTIVES IN PHYSICS 184
- CONTRIBUTIONS OF WOMEN IN QUANTUM SCIENCE 185
- ADDRESSING GENDER BIAS IN STEM FIELDS 186

XLIV. QUANTUM MECHANICS AND SOCIAL SCIENCE 187
- SOCIOLOGICAL IMPACT OF QUANTUM DISCOVERIES 188
- QUANTUM MECHANICS IN SOCIETAL DECISION-MAKING 189
- INTERDISCIPLINARY RESEARCH AND COLLABORATION 190

XLV. QUANTUM MECHANICS AND ETHICS 191
- ETHICAL IMPLICATIONS OF QUANTUM TECHNOLOGIES 192
- RESPONSIBILITY IN SCIENTIFIC RESEARCH 193
- ETHICAL EDUCATION FOR QUANTUM SCIENTISTS 194

XLVI. QUANTUM MECHANICS AND LANGUAGE 195
- TERMINOLOGY AND CONCEPTUAL UNDERSTANDING 196
- LANGUAGE AS A TOOL FOR TEACHING QUANTUM PHYSICS 197
- COMMUNICATION OF QUANTUM IDEAS TO THE PUBLIC 198

XLVII. QUANTUM MECHANICS AND PSYCHOLOGY 199
Cognitive Approaches to Quantum Concepts 200
Psychological Impact of Quantum Discoveries 201
Quantum Mechanics in Psychological Theory 202

XLVIII. QUANTUM MECHANICS AND THE ARTS 203
Intersections Between Quantum Physics and Artistic Expression 204
Art as a Medium for Explaining Quantum Ideas 205
Collaborations Between Artists and Physicists 206

XLIX. QUANTUM MECHANICS AND INNOVATION 207
Driving Technological Advances 208
Quantum Innovation Ecosystems 209
Fostering Creativity in Quantum Research 210

L. QUANTUM MECHANICS AND THE FUTURE 211
Emerging Technologies and Their Impact 212
Quantum Mechanics in Future Societies 213
Speculations on the Evolution of Quantum Science 214

LI. CONCLUSION 215
Summary of Key Points Discussed 216
Reflection on the Role of Quantum Mechanics in Science and Technology 217
Future Prospects and Directions in Quantum Research 218

BIBLIOGRAPHY 219

I. Introduction to Quantum Mechanics

The foundational pillar of contemporary physics, quantum mechanics, ushers in an era that defies typical comprehension. It was brought to life by forefront scientists such as Max Planck, Niels Bohr, and Werner Heisenberg. This branch of science plunges into the minuscule universe to reveal essential principles that command the demeanor of both matter and energy. Quantum mechanics is primarily concerned with phenomena such as superposition — a condition enabling particles to be in various states at once — and quantum entanglement which establishes an inexplicable link among particles irrespective of the space between them. Together with concepts like wave-particle duality and the wave function, these elements lay down the groundwork for quantum mechanics, providing fresh insight into the universe's complexities. Through explaining these principal notions through examples and visuals, novices may start unraveling quantum mechanics' intricacies and recognizing its pivotal role in molding our perception of reality.

Definition of Quantum Mechanics

The fundamental theory of quantum mechanics stands as a pivotal element in the architecture of contemporary physics, overturning conventional perspectives on existence and revealing deep insights into the realm at a minuscule scale. Figures such as Max Planck, Niels Bohr, and Werner Heisenberg pioneered this field by investigating how particles and waves operate at an elemental level, with a focus on phenomena like superposition, entanglement in quantum context, and duality between wave-particle aspects. Superposition permits the simultaneous occurrence of multiple states within particles. Entanglement ensures instantaneous linkage among particles despite extensive separations contradistinguishing age-old perceptions about space-time fabric. The concept concerning wave-particle duality illuminates the ambidextrous attributes of quantum bodies that demonstrate both corpuscle-like and undulatory behaviors when observed closely. These essential principles constitute the complex framework of quantum mechanics altering our comprehension toward cosmos intricacies thereby laying foundational stones for revolutionary breakthroughs across technological fields and theoretical paradigms in physics.

Importance in Modern Physics

Modern physics is fundamentally underpinned by the discipline of quantum mechanics, transforming our perceptions of the universe's basic constituents. Its importance goes beyond theoretical ideas, influencing practical developments that are at the forefront of avant-garde technologies and scientific progress. The groundbreaking efforts of innovators such as Max Planck, Niels Bohr, and Werner Heisenberg have indefinitely modified the field of physics, introducing an era dominated by quantum phenomena that defy traditional principles. Notions like super positioning, entanglement in quantum realms, and the dual nature of particles and waves do more than just alter our grasp on reality; they provide a critical foundation for transformative tech advancements including quantum computation and sophisticated encryption methods. To dive into future breakthroughs within science and technological spheres necessitates embracing quantum mechanics' intricacies. This effort will drive us towards uncharted territories in understanding our cosmos better—foreseeing a realm filled with endless opportunities for reimagining what we know about our world's architecture.

Overview of Essay Structure

In dissecting the essay blueprint for 'Deciphering Quantum Mechanics: Key Concepts Explained for Beginners,' it's pivotal to unfold a detailed exposition that ensnares and enlightens the intended readership adeptly. The commencement ought to delineate quantum mechanics and spotlight its paramountcy in contemporary physics, charting its ancestral lineage back to vanguard scientists such as Planck and Bohr. Advancing into the cardinal notions segment, an explication of fundamental principles like superposition, quantum entanglement, and wave-particle duality is indispensable for comprehending the complexities within quantum doctrine. Vivid instances and depictions are imperative for elucidating these multifaceted notions to novices. Shifting towards theoretical dogmas, an equitable depiction of core formulae akin to Schrödinger's equation, merged with varied exegeses such as Copenhagen's and Everett's interpretations will render a panoramic view whilst eschewing numerical conundrums. Illustrating the pragmatic employments of quantum mechanics in apparatuses like quantum computing and cryptography accentuates its tangible influence. Impartial scrutiny should openly tackle quandaries in grasping and explicating quantum mechanics, broadening discussions to criticisms also wrangles prevalent within this sphere. A robust denouement should amalgamate essential takeaways while contemplating future vistas of quantum mechanics in scientific plus technological progressions ensuring a cogent instructive composition that brightens further piquing curiosities amongst aficionados.

II. Historical Context

Delving into the backdrop of quantum mechanics unveils an epoch characterized by the transformative inputs from notables such as Werner Heisenberg, Niels Bohr, and Max Planck. Originated in the dawn of the 20th century, quantum theory instigated a shift in scientific paradigms, upending traditional Newtonian doctrines while fostering a refined scrutiny within particles' minute boundaries. Such innovators constituted the groundwork for principles including entanglement, quantum states, and Heisenberg's Uncertainty Principle that today shape our grasp on reality and cosmos comprehensions. Their pioneering endeavors didn't just birth quantum mechanics but also smoothed paths towards technological leaps in encryption alongside computing sectors; thus morphing industry landscapes along with remolding collective insights on universe's interwoven essence. The evolution path of quantum mechanics epitomizes human creativity besides an unyielding quest for unraveling prime enigmas of the quantal sphere catered to both novices and connoisseurs alike.

Early 20th Century Physics

At the dawn of the 20th century, a significant transformation in physics took place with quantum mechanics stepping onto the scene, posing challenges to traditional theories and altering our grasp of tiny particles. European scholars such as Max Planck, Niels Bohr, and Werner Heisenberg were at the forefront of this revolution, bringing forth radical ideas that still influence today's physics and technological advancements. The establishment of fundamental notions like states of quantum, intertwining particles, Heisenberg's Uncertainty Principle, and dual aspects of waves and particles constructed the groundwork for this novel framework. These notions faced initial doubts but found real-world uses in realms like computation, telecommunication methods, through advanced speed in calculations and heightened protection via concepts such as super positioning fluids within intricacies branching further than just expanding scientific inquiry borders; it incited meditations on existence's essence alongside considerations about how entangled quantum figures might be. Venturing into early 20th-century physics lays down an elementary platform for novices eager to unravel quantum mechanics' complications by offering dense historical insights alongside a conceptual base conducive to broader investigation.

Key Scientists and Their Contributions

In the realm of quantum mechanics, essential contributions have been made by leading intellectuals such as Max Planck, Niels Bohr, and Werner Heisenberg, whose pioneering efforts have significantly molded contemporary physics. The initiation of energy quantization and the emergence of quantum theory by Planck laid crucial groundwork for further inquiries in this domain. Conversely, Bohr's investigations into the atomic framework and his formulation of the Copenhagen Interpretation questioned conventional cosmic perspectives. Introducing a novel degree of unpredictability at the quantum dimension through Heisenberg's Uncertainty Principle undermined determinist standpoints. Accompanied by figures like Dirac and Feynman, these academicians have facilitated technological progressions and deepened our philosophical grasp on existence revealing how quintessentially interconnected quantum elements are with complex undercurrents defining our cosmos's essence. Their seminal works endure as catalysts propelling investigative endeavors within sectors such as quantum computing and communication technologies, transcending limitations while revolutionizing our empirical worldview.

Milestones in Quantum Theory Development

Unraveling the key moments in the progression of quantum theory reveals an intriguing voyage filled with significant scientific discoveries and shifts in paradigms. The transition from classical mechanics to quantum physics was highlighted by essential contributions from pioneers such as Planck, Bohr, and Heisenberg, who overturned traditional beliefs with their revolutionary ideas. Principal landmarks comprise the formation of quantum field theory by Dirac and Feynman, which elucidated on particle interactions and forces present in nature, alongside the Copenhagen Interpretation that suggested outcomes based on probability within the quantum domain. These pivotal points transformed contemporary physics and technological applications, creating pathways for progressions in areas like computing and cryptography. Grasping these critical junctures is vital for novices venturing into the intricate world of quantum mechanics, offering a base to apprehend reality's interconnected essence along with universal mysteries through clear-eyed wonderment and inquiry.

III. Wave-Particle Duality

The principle of wave-particle duality, a key idea in quantum mechanics, overturns traditional viewpoints by unveiling the dual aspect of particles as both waves and particles. Demonstrated vividly in the renowned double-slit experiment, this duality reveals the baffling actions of quantum entities, where particles demonstrate characteristics akin to waves and conversely. Grasping this concept is vital for understanding the intertwined essence of quantum systems and their non-traditional operational rules. Diving into wave-particle duality aids novices in discovering the complex link between particles and waves, thus facilitating a more profound knowledge of quantum mechanics. This notion not only accentuates the cryptic attributes of quantum beings but also stresses on revising orthodox physics theories, laying groundwork for further investigations and discoveries within the domain of quantum mechanics.

Concept of Duality

In quantum mechanics, the duality principle showcases the astonishing scenario where particles display characteristics of both waves and particles at once, posing a challenge to classical physics principles. Embodied by the notable double-slit experiment, wave-particle duality highlights the inherent uncertainty that lies at core of quantum beings. This dual nature obscures the old distinctions drawn between waves and particles, prompting a reconsideration of our grasp on reality and what constitutes existence. Viewed through duality's prism, quantum mechanics unfolds a complex web within the microscopic universe, hinting at an underlying layer of reality unseen to our typical observations. Investigating quantum mechanics' notion of duality opens doors to decoding the universe's intricacies, encouraging rookies to ponder over this scientific field's bewildering array of potential outcomes and ambiguities that define its peculiar domain.

Experiments Demonstrating Duality

The phenomenon of duality within quantum mechanics stands as a puzzling contradiction to our traditional perceptions, asserting that entities can exist as both waves and particles. The double-slit experiment among others showcases this confounding attribute where entities exhibit characteristics of both wave and particle based on observation methods. This key principle not only challenges the standard view of reality but also highlights the deep connections between quantum elements. Through such experiments, newcomers are able to understand the complex dynamics at play in quantum phenomena and appreciate their significant impact on science and technological advancements. As newcomers delve into these basic experiments showing duality, they step into the mysterious world of quantum mechanics, prompted to rethink their views on universe's structure and matter's core essence, thereby gaining an advanced understanding of this groundbreaking domain.

Implications for Understanding Matter

The deep effects of quantum mechanics on our comprehension of matter are unfolded, revealing how particles and fields are deeply interconnected. By bringing to light ideas such as quantum entanglement and the dual nature of wave-particle, it muddies the waters for classical interpretations of what matter is about. The manuscript delves into how both the Principle of Uncertainty and the Effect observed by participants point out matter's unstable and unforeseeable character when scrutinized at a quantum scale, leading to a revolutionary change in our outlook on universe's basic components. It motivates novices to question old perceptions regarding matter's conduct, setting up foundations for grasping reality's complex weave more thoroughly. These implications drawn from quantum notions for unraveling material substance's core underscore an adaptable and non-rigid mentality needed while diving into microscopic realm's enigmas.

IV. The Quantum State and Superposition

Delving into the complex domain of quantum mechanics reveals the pivotal role played by the quantum state and superposition, challenging our conventional beliefs about reality. The entirety of a quantum system's information is captured in its quantum state, highlighting its inherent uncertainties and probabilities. Superposition introduces an enigmatic contradiction within this framework, allowing particles to coexist in various states at once until they are measured, starkly opposing classical reasoning. Such phenomena don't just underscore the inherently unpredictable characteristics of quantum mechanics but also herald groundbreaking progress in areas such as quantum computing and cryptography. By understanding the crux of superposition and the quantum state, novices can penetrate deeper into the enthralling intricacies that define quantum mechanics, thereby unveiling its capacity to reshape our future technological landscape and scientific modalities. With lucid explanations accompanied by illustrative instances, individuals can journey through these fundamental concepts' theoretical maze—preparing for an enriched comprehension of the atomic cosmos alongside its staggering influence on how we interpret both existence and cosmological vastness.

Explanation of Quantum States

Grasping the intricacies of quantum mechanics fundamentally relies on a deep comprehension of quantum states. These states outline the potential configurations in which a quantum system might find itself, highlighting the inherent ambiguity and overlap characteristic of the quantum domain. Described through wave functions - mathematical representations that define outcome probabilities upon observation - these states capture complexity within their scope. Entanglement, where particle attributes are interconnected regardless of spatial separation, disrupts traditional ideas about space-related properties. This entwined nature brings to light the notion of wave-particle duality, illustrating how particles can display characteristics akin to both waves and distinct entities based on how observations are conducted. Understanding quantum states offers insights into the probabilistic interactions and determinants central to mastering fundamental tenets of quantum mechanics, thus unlocking new perspectives on this enigmatic yet captivating universe's essence.

Principle of Superposition

As a cornerstone idea in the study of quantum mechanics, the Principle of Superposition fundamentally challenges our traditional views on reality and opens new doors for novices venturing into this intricate domain. It posits that until an observation is made, a quantum system remains in every possible state at once, representing not just one certain condition but rather a multitude of potentialities. This notion, brought to light by figures such as Schrödinger, underscores the profound uncertainty and complex relationships within the quantum universe, illustrating how particles can display characteristics akin to waves and exist across diverse states all at once. Grasping superposition is crucial for understanding the strange behaviors of quantum objects and sets the stage for delving into phenomena like wave-particle duality and quantum entanglement. Through mastering superposition, those new to this field can decode the elaborate structure of quantum mechanics and recognize the dynamic interplay that defines our cosmos.

Thought Experiments and Illustrations

Delving into the domain of quantum mechanics for novices, illustrations and mental exercises are crucial in making intricate ideas both clear and profound. Mental exercises such as Wigner's friend or Schrödinger's cat act as cerebral tools to disrupt traditional comprehension and stimulate in-depth consideration regarding quantum phenomena's essence. By introducing imaginary situations that clash with classical logic, these mental ventures encourage learners to struggle with unpredictability and the intrinsic linkage found within quantum mechanics. Visual aids like visualizations of quantum entanglement or the double-slit experiment help in embodying abstract notions including wave-particle duality and superposition. Via visual representations and thought adventures, newcomers can interact with principal quantum theories in a manner that is both insightful and perceptible, facilitating an advanced grasp on the implications of the quantum universe for contemporary physics and technology advancements.

V. Quantum Entanglement

Grasping the essence of quantum entanglement is essential for rookies to unravel the enigmas of quantum mechanics when venturing into the intricacies within the realm of quanta. Termed as "spooky action at a distance" by Einstein, quantum entanglement signifies an occurrence where particles remain deeply bonded irrespective of their spatial separation. This intertwined state leads to correlated attributes, like spin or polarization, staying connected despite being miles apart, hinting at some form of unseen linkage that transcends traditional comprehension. Investigating this phenomenon opens doors to understanding non-locality and possible interconnectedness lurking in quantum frameworks. Through clarifying such tangled connections, novices might start to decipher complexities inherent in quantum mechanics and its ramifications for our perception of actuality and cosmos existence.

Definition and Basic Principles

For beginners to unravel the complex domain of quantum mechanics, it's crucial to first understand its definition and foundational tenets. At modern physics' core, quantum mechanics explores the actions of particles at a subatomic level and the universe's elemental functions. This field was forged by the collective genius of individuals like Max Planck, Niels Bohr, and Werner Heisenberg, pushing beyond traditional beliefs to introduce a revolutionary perspective on reality. Fundamental concepts such as superposition, entanglement in quantum realms, duality between waves and particles, along with wave function theories are essential for building an understanding of quantum theory; this often requires clear visual aids and concrete examples for better comprehension. These ideas introduce layers of unpredictability and interconnected phenomena that lead into deeper theoretical discussions about entities like Schrödinger's equation as well as differing interpretations including those from Copenhagen to Many-Worlds perspectives. Diving into the mystifying world of quantum mechanics allows newcomers to appreciate how it significantly influences areas like quantum computing and encryption technology while encouraging critical debates over its complexities, critiques faced within this intriguing area of study, also speculating on its role in carving out new scientific explorations and technological innovations across various fields.

EPR Paradox and Bell's Theorem

In 1935, Einstein, along with Podolsky and Rosen, introduced the EPR Paradox, casting doubts on quantum mechanics by pointing out supposed intrinsic uncertainties within the theory. This paradox argued that a comprehensive understanding of physical reality was beyond quantum mechanics' grasp due to assumed concealed variables. Further exploring these notions, John Bell came forward in 1964 with Bell's Theorem, which sought to verify these hidden variables via experimental tests with entangled particles. Through establishing criteria for evaluating if quantum mechanics or local hidden variables provided a more accurate account of correlations in entangled systems, Bell's experiments leaned towards supporting the probabilistic essence of quantum theories over fixed outcomes. Digesting the EPR Paradox and Bell's Theorem is vital for anyone new to diving into the depths of quantum mechanics debates; it serves as a complex gateway to understanding critical arguments at play within this field.

Implications for Information Theory

In the sphere of quantum mechanics, the repercussions for Information Theory present captivating perspectives on the core aspect of how information is processed and conveyed. The ability of quantum states to be in multiple classical states simultaneously disrupts conventional methods of binary encoding, heralding advancements in computing efficiencies through novel algorithms. Quantum entanglement introduces groundbreaking prospects for secure messaging and the phenomenon of quantum teleportation by creating indissoluble links between particles irrespective of their separation, thus transforming our understanding of information dissemination. Moreover, the aspects like wave-particle duality alongside the Uncertainty Principle introduce complexities into how information is encoded and decoded, necessitating a revision in deterministic approaches to Information Theory. Engaging with these phenomena allows novices to comprehend that quantum mechanics does not just bring about a paradigm shift in computing but also reshapes vital groundwork within Information Theory, thereby setting a course towards future breakthroughs in data handling and exchange.

VI. The Uncertainty Principle

Positioned as a fundamental keystone within the realm of quantum mechanics, the Uncertainty Principle directly confronts traditional beliefs surrounding predictability and certainty in our universe. Introduced by Werner Heisenberg back in 1927, this principle posits that attempting to measure a particle's position with high accuracy inherently decreases the precision with which its momentum can be known, and the other way around. It suggests an inherent element of randomness and unpredictability at the very heart of quantum behavior, highlighting our knowledge limitations and how deeply interconnected quantum entities are. For novices venturing into the domain of quantum mechanics, comprehending the Uncertainty Principle is crucial because it embodies the quintessential unpredictability characterizing particle dynamics. By understanding this principle, learners are equipped to decode complex aspects of quantum reality better and explore diverse facets of such microscopic phenomena from an enriched vantage point.

Heisenberg's Contribution

The advancements made by Heisenberg in the domain of quantum mechanics mark a crucial turning point, transforming our comprehension of entities at a microscopic scale. The introduction of the Uncertainty Principle by Werner Heisenberg disrupted the conventional principles laid out by Newtonian physics, highlighting that there are inherent constraints to measuring both the velocity and location of a particle simultaneously with perfect accuracy. Such an innovative principle has drastically changed our perspective towards subatomic particle behavior, underlining quantum phenomena's fundamentally stochastic characteristic. Not merely did Heisenberg's endeavors lead to a scientific paradigm shift but they also set the stage for embracing notions of indeterminacy within physics. Through revealing observation limits at the minuscule level, his revelations provided groundwork for subsequent progress in quantum mechanics which influences areas like cryptographic techniques and computing based on quantum theory. The broader implications of this principle break through disciplinary barriers, showcasing how entwined quantum elements challenge age-old deterministic outlooks—thereby underpinning it as critical learning material for those exploring quantum mechanics' intricacies.

Measurement and Precision Limits

Diving into the complex world of quantum mechanics unveils the pivotal notion of 'Measurement and Precision Limits,' a cornerstone aspect that shapes the core of this baffling domain. At the heart of quantum mechanics thrives its fundamental uncertainty, symbolized by Heisenberg's Uncertainty Principle. This principle imposes limitations on how accurately pairs of properties such as position and momentum can be measured at once. It throws traditional beliefs in determinism into question, highlighting instead the probabilistic essence inherent in quantum occurrences. Additionally, the nuanced relationship between observation and measurement in quantum realms emphasizes a fragile equilibrium between accuracy and ambiguity. For novices disentangling the conundrums of quantum mechanics, wrestling with these intrinsic measurement constraints marks an essential phase in understanding the intricate exchanges among quantum events and grasping reality's elusive character.

Philosophical Implications

When exploring the philosophical ramifications of quantum mechanics, an essential transformation in how we perceive reality takes place, putting traditional views on determinism and causality to the test. The complexity and unpredictability found within quantum occurrences urge us to ponder over the concept of free will and consciousness, leading to a meditation on how observation impacts our reality. Ideas such as entanglement and superposition in the quantum realm obscure the lines between those observing and what is observed, hinting at a complex interplay between human awareness and quantum intricacies. Quantum mechanics' reach goes beyond just scientific discourse, encouraging a reassessment of our position within a universe where notions of uncertainty, connectivity, and observer effect intertwine to reshape our philosophical outlooks. For novices delving into quantum mechanics' depths, engaging with its philosophical questions opens doors to profound reflection and admiration for the enigmas anchoring our cosmos comprehension.

VII. The Wave Function

As a fundamental element in the realm of quantum mechanics, the Wave Function crucially encapsulates the probabilistic essence inherent to quantum systems and reflects the condition of a system or particle exhibiting wave-like attributes. This representation through mathematics affords predictions regarding how particles will act, shedding light on their location, momentum, and energy dynamics. To come to grips with the wave function necessitates wrestling with the sophisticated idea that entities display characteristics of both waves and particles at once, overturning conventional perceptions of what constitutes reality. For novices making sense of quantum mechanics' intricacies, an understanding of the wave function is critical for unlocking concepts such as superposition and entanglement among quantum entities. The role played by this mathematical tool not only expands our insight into countless uncertainties and capabilities within physics but also lays foundational stones for imminent breakthroughs across scientific research fields.

Role in Quantum Mechanics

Within the sphere of quantum mechanics, holds a crucial position in altering our comprehension of how everything in the universe is interconnected and transforming traditional beliefs about reality. Phenomena such as quantum entanglement and superposition make contest notions that are deterministic and add elements of unpredictability into the essence of what exists. Exploring the intricacies of causes us to rethink our views on space and time, leading to consideration over basic laws that manage entities at a quantum level. The impact of goes past just theoretical discussions; it infiltrates practical uses like in quantum computing and cryptography, where its effect leads to major advances in technology and safeguarding information. Grasping the subtle aspects of not only widens our grasp on scientific endeavors but also promotes an adaptable and broad-minded stance towards solving mysteries within quantum mechanics, creating paths for remarkable discoveries and deep philosophical inquiry.

Interpretation of the Wave Function

Delving into the interpretation of the quantum mechanics wave function is vital for grasping how particles behave within the quantum domain. This function embodies the probability-based characteristics of quantum entities, shedding light on their conditions and actions. There are several schools of thought trying to demystify what this wave function means and its importance, sparking discussions about how it influences results when measurements are made. The Copenhagen school argues that a definite state emerges from the collapse of the wave function during observation, highlighting observation's impact in quantum realms. In contrast stands the Everett theory, which proposes a proliferating multiverse where every potential outcome exists simultaneously, disputing traditional perceptions of reality's nature. Venturing through these theories equips novices with an essential grasp on quantum mechanics' intricacies, setting up a gateway for further inquiries into both theoretical aspects and tangible implementations of quantum phenomena.

Probability and Measurement

Within the field of quantum mechanics, measurement and probability emerge as fundamental concepts that challenge conventional ideas of determinism. Quantum states reside in superposition, meaning outcomes are determined by chance until observed, a theory underscored by the Copenhagen Interpretation. The principle of uncertainty isn't limited to particle behavior but also encompasses the act of measuring itself, which is inherently random. Through quantum mechanics' perspective, observing phenomena becomes an intricate dance between probabilities and uncertainties, underscoring how interconnected reality truly is while highlighting classical certainty's constraints. The integration of quantum principles into areas like quantum computing and cryptography showcases probability's critical role in driving forward technological breakthroughs, underlining the importance of mastering uncertainty for achieving revolutionary advancements in science and technology.

VIII. The Schrödinger Equation

To grasp the intricacies of quantum mechanics, it's crucial to understand the central role played by the Schrödinger equation as an underpinning theoretical structure. Introduced by Erwin Schrödinger in 1925, this pivotal formula represents the progression of a wave function within a quantum framework across time, offering a calculative portrayal of particle dynamics at microscopic scales. At initial observation, while tricky to comprehend, its core functionality is aimed at forecasting potential outcome probabilities for specific systems. Engaging with the Schrödinger equation allows novices to acquire basic insights into quantum mechanics' fundamental concepts such as wave-particle duality and superposition necessary for navigating through quantum domains. This formula acts as an entry point for investigating quantum manifestations and apprehending how particles act following probabilistic norms thus playing a critical role in unraveling mysteries that permeate the field of quantum mechanics for budding enthusiasts.

Introduction to the Equation

Diving into the realm of quantum mechanics fundamentally revolves around understanding the core equation that serves as its backbone. The pivotal Schrödinger equation, essential within the domain of quantum mechanics, inaugurates an exploration into quantum system dynamics. Erwin Schrödinger introduced it as a particle wave equation, thus creating a conduit between classical physics and its quantum counterpart and providing probabilistic views on microscopic particle behavior. Beginners in this field are met with a combination of mathematical grace and deep conceptual insights, sculpting their perspective on phenomena such as entanglement and superposition intrinsic to quantum realms. Through engaging with the Schrödinger equation, novices initiate their odyssey towards demystifying quantum mechanics' secrets thereby fostering an enriched comprehension of both its empirical applications and theoretical fabric sewn by historical vanguards in science.

Time-Dependent vs. Time-Independent Forms

For novices unraveling the complexities of quantum mechanics, it's crucial to grasp the difference between time-dependent and time-independent versions. Time-dependent wave functions alter over time in line with Schrödinger's equation, showcasing the fluctuating character of quantum entities. This version is critical for depicting occurrences where a system's state changes through time, shedding light on aspects like particle exchanges and disintegration. Conversely, time-independent wave functions depict unchanging states with invariant energy levels, facilitating the determination of energy eigenvalues and likelihoods devoid of temporal change contemplation. These versions lay down an essential framework in quantum mechanics, enabling a thorough examination of systems across varying periods. Acquiring knowledge about both forms furnishes novices with necessary capabilities to navigate through numerous quantum events, establishing a robust groundwork for delving into quantum mechanics' complex realm.

Conceptual Understanding Without Math

Delving into quantum mechanics, especially for novices aiming to acquire a basic conceptual grasp sans the entanglement in elaborate math, necessitates an understanding of the core principles anchoring this complicated physics domain. Although mathematical formulae such as the Schrödinger equation serve as crucial cornerstones for formalizing quantum mechanics, attaining a subtle appreciation of principal notions like superposition, quantum entanglement, and the dual nature of particles and waves is plausible via intuitive elucidations and comparisons drawn from everyday life. By simplifying these essential concepts into more understandable bits, one can recognize the complex interrelations and inherent unpredictability characteristic of quantum happenings without drowning in stringent mathematical details. This strategy aids not just in developing a conceptual scaffold for grasping quantum mechanics but also ignites an interest that might propel further ventures into unraveling its intricacies and discerning its repercussions on contemporary technology along with scientific reasoning.

IX. Quantum Tunneling

Quantum tunneling, an enthralling concept within quantum mechanics, defies our conventional grasp of obstacles and the mobility of particles. When a particle maneuvers through a potential energy barrier that it traditionally shouldn't surpass, this is when quantum tunneling manifests. This somewhat astonishing act can be showcased by electrons' actions in semiconductors, which are crucial for powering up devices essential for contemporary tech. By sinking into the depths of quantum tunneling theories, novices may unlock the complex essence of quantum mechanics alongside its everyday utility. Fathoming how particles manage to burrow through barriers paves a path toward delving into the weave and erratic nature of the quantum domain, casting illumination upon subatomic enigmas. In realizing the theory behind quantum tunneling, learners set off on an expedition to discover the intricacies lying at heart of quantum mechanics—unearthing its capacity to transform diverse scientific and technological arenas profoundly.

Phenomenon Explanation

Delving into the mysteries of quantum mechanics ushers in a transformative alteration in how beginners apprehend both reality and the cosmos. The notions of quantum states, entanglement, and the Principle of Uncertainty dismantle conventional perspectives, urging a reassessment of core principles. Such critical revelations lay the groundwork enabling individuals to navigate through the intricate realm of quantum physics. Through scrutinizing wave-particle duality, participants are encouraged to question particle nature and quantum entities' mutual connections. Contributions from prominent figures like Max Planck, Niels Bohr, and Werner Heisenberg illuminate this discourse's historical backdrop, accentuating quantum theory's groundbreaking influence on contemporary science and technological innovation. This delineation underscores how quantum mechanics heralds an era of exploring scientific frontiers beyond established boundaries – ushering new horizons in computation, encryption practices, and philosophical reflection.

Applications in Electronics

In the realm of electronics, the transformative essence of quantum mechanics becomes evident, ushering in an era filled with advanced technological breakthroughs. Through superposition, electronic devices see a surge in efficiency and computational prowess. Entanglement, crucial to quantum mechanics' core, underpins the development of highly secure quantum communication networks, enhancing encryption methods within electronic domains. The principle of wave-particle duality serves as a guidepost for innovating versatile components in next-gen electronics. Leveraging these tenets from both quantum computing and cryptography contributes to accelerated processing rates alongside unparalleled data protection measures. The adoption of quantum mechanics by those in research and engineering heralds a myriad of potential innovations while steering technology's future direction. This fusion between quintessential quantum theories and practical applications marks a significant milestone for modern electronic advancements--signaling unmatched progress and exploratory ventures into new technological frontiers.

Conceptual Challenges

Beginners diving into quantum mechanics grapple with a significant paradigm shift in their grasp of reality and cosmos understanding. Fundamental theories such as wave-particle duality, superposition, and quantum entanglement upend conventional wisdom, demanding an essential reassessment of how we view our surroundings. The complex character of the wave function alongside the consequences stemming from observing quantum occurrences add layers that elude straightforward rational thinking. Foundations driving quantum mechanics, for instance, the Schrödinger equation along with diverse interpretations like Copenhagen and many-worlds explanations enrich both complexity and philosophical nuance to this field. Notwithstanding its practical deployment in cutting-edge innovations like cryptography and quantum computing, navigating through conceptual hurdles within quantum mechanics poses an engaging yet formidable pursuit for novices eager to unlock the enigmas enveloping the realm of quanta.

X. Quantum Computing

Leveraging quantum mechanics principles, quantum computing marks a significant leap forward in technological innovation, reshaping computation methods dramatically. By exploiting superposition and entanglement characteristics, this advanced computational model can undertake vast calculations at speeds unattainable by conventional computers. Quantum bits or qubits' manipulation allows these machines to exist in manifold states at once, enabling the examination of multiple scenarios simultaneously and thus delivering extraordinary computational strength. Such groundbreaking technology paves the way for tackling intricate issues across various domains like cryptography and pharmaceuticals with unmatched speed and efficiency. As quantum computing progresses, its adoption across different industries is set to transform our methodologies concerning data evaluation, optimization exercises, and complex problem resolutions—ushering us into an era characterized by computational abilities that surpass previous barriers.

Basics of Quantum Computers

Interpreting the elementary concepts of quantum computing unveils a sphere of tech advancements set to transform the field of computation. At its heart, superposition serves as a foundational principle, with qubits capable of existing in various states at once, thereby significantly boosting computational capacity. Quantum entanglement magnifies this effect by intertwining qubits in ways that defy traditional physics principles, facilitating instantaneous data transfer and solving intricate challenges. Such quantum events disrupt the normative binary computing models, paving the way for potential calculations at an exponentially faster rate. Grasping basic notions such as wave-particle duality and wave function is essential to comprehending the singular processing abilities endowed by quantum computing. As developments in quantum technology advance from theory towards tangible applications, we stand on the brink of unlocking remarkable levels of computational power and altering technological paradigms within our digital age.

Qubits and Quantum Supremacy

Comprehending the basics of qubits alongside their significance in realizing quantum dominance is crucial for novices unraveling the mysteries of quantum mechanics. Qubits, which serve as the foundational elements of quantum computing, stand apart from traditional bits through harnessing superposition and entanglement's phenomena to execute intricate computations at speeds without rivals. Such a critical variance empowers quantum systems to eclipse conventional machines in specific tasks, an achievement referred to as quantum supremacy. Quantum supremacy marks the juncture where a quantum device can undertake operations beyond what classical mechanisms can handle, illustrating a potential upheaval in computational force. Through understanding both the importance of qubits and the concept of quantum supremacy, beginners are ushered into an era brimming with ground-breaking opportunities and a reimagined vista on computer science's horizon, thereby elevating them into spheres filled with innovative exploration and reshaping technological progress along scientific discoveries.

Potential Impact on Society

Quantum mechanics' impact on societal structures is intricate and deep, presenting a new perspective that touches on many elements of everyday existence. Technologies rooted in quantum theory, like computing and cryptography within the quantum realm, could drastically alter sectors such as telecommunications, health services, and ecological conservation. The creation of materials and sensors based on quantum principles is poised to enhance processes related to waste management significantly, optimize energy use, and bolster efforts against climate alterations. The ethical dilemmas presented by the application of quantum technology—particularly concerning individual privacy, security measures, and the balance of global authority—underscore the urgency for all-encompassing regulatory systems to oversee its implementation. By tackling these moral issues while promoting cross-border collaboration, humanity stands at the threshold of fully leveraging quantum breakthroughs for collective betterment—a step towards ensuring a future that's both greener and safer.

XI. Quantum Cryptography

Quantum cryptography, an essential application of quantum mechanics, transforms data protection by employing quantum characteristics to establish indecipherable encryption. Through adopting frameworks like quantum key distribution, it guarantees communication pathways secure from intrusion attempts. This advanced mechanism relies on spectacles such as quantum entanglement, presenting a considerable advancement in cybersecurity. The deployment of quantum cryptography demonstrates the real-world applications of quantum mechanics outside academic exploration, illustrating its revolutionary capability in protecting delicate data today. As newcomers navigate through the complexities of deciphering quantum mechanics, grasping the nuances of quantum cryptography showcases how basic concepts can be applied to overcome current obstacles, emphasizing the concrete influence of quantum theories on contemporary technology and encryption techniques.

Principles of Quantum Encryption

In secure communication protocols, the tenets of quantum encoding usher in an era marked by a radical alteration, utilizing the innate erratic nature and linkage characteristics inherent in quantum physics to protect confidential data. Quantum encryption is predicated on the superposition effect, allowing qubits to be in numerous states at once, thereby rendering any intercepted information ambiguous to unauthorized listeners. This novel cryptographical method benefits from quantum entanglement, whereby particles that are spatially separated can immediately reflect one another's conditions across great distances, ensuring a level of data protection surpassing that offered by traditional cryptographic techniques. Through adopting these elementary notions from quantum mechanics, encryption technology undergoes a transformational shift—providing security for information at rates never before seen as vital in this age dominated by data reliance. The inclusion of quantum principles within encryption schemes carves out routes for sophisticated digital safeguard strategies while underscoring the significant role played by quantum mechanics within contemporary technological progression.

Quantum Key Distribution (QKD)

At the pinnacle of secure communication technologies lies Quantum Key Distribution (QKD), which capitalizes on quantum mechanics' principles for achieving unmatched secrecy in data exchanges. Through harnessing quantum occurrences such as superposition and entanglement, QKD facilitates the creation of encryption keys that are impossible to crack, thereby ensuring secure communication pathways. This cutting-edge method drastically alters the landscape of cybersecurity by providing a robust solution for safeguarding confidential information against nefarious infiltrations. The complex characteristic of quantum entanglement enables the generation of encryption keys whose connection is inherent and resistant to eavesdropping, establishing an unprecedented standard in data protection. As novices navigate through quantum mechanics' domain, gaining insights into QKD illuminates how quantum theory's practical aspects are pivotal in defending information within our digital and extensively connected milieu. Delving into QKD not merely illustrates the intersection between abstract quantum ideologies and tangible technological deployments but also highlights how profoundly quantum mechanics influences envisioning secure communication's future contours.

Future of Secure Communication

In the landscape of future secured messaging, quantum physics stands as a trailblazing power transforming online interactions. Based on quantum entanglement and superposition theories, quantum cryptography presents indestructible coding methods that aim to overhaul information protection. The creation of communication networks utilizing quantum methods such as quantum key distribution signifies an essential move towards defending confidential data from digital risks. With the progression of quantum innovations, the emergence of a worldwide quantum internet becomes imminent, envisaging protected and efficient channels for correspondence that overcome existing barriers. Yet, as we voyage towards this future powered by quantum mechanics, it's vital to meticulously deliberate over ethical issues related to privacy, security measures, and international supremacy dynamics to guarantee cautious application and use of these ground-breaking advancements. By acknowledging the limitless prospects offered by quantum communications technology, we are poised at the dawn of a new era where unparalleled levels of safety and interconnectivity redefine our digital existence for upcoming generations.

XII. Quantum Teleportation

Within the intriguing quantum mechanics field, quantum teleportation captures our imagination by allowing for the immediate transfer of quantum information over large distances without the need for physical movement. This occurrence, grounded in the notions of quantum entanglement and superposition, upends typical ideas regarding communication and space constraints. The instantaneous correlation of states between a duo of entangled particles during teleportation carries information surpassing old-fashioned limits. Quantum teleportation harbors significant promise for transforming areas such as secure messaging and quantum computing, heralding a new era in tech advancements. Diving into quantum teleportation's complexities offers novices a peek into the puzzling domain of quantum mechanics, showcasing how particle interconnectivity and the transformative essence of quantum events are intertwined. As an essential principle for unraveling the mysteries within quantum mechanics, teleportation highlights novel opportunities unlocked when exceeding conventional barriers, forging an innovative framework for technological breakthroughs and probes into the enigmatic depths of the quantum universe.

Theoretical Framework

Within the theoretical construct of quantum mechanics, a complex lattice of theories and notions weaves together to upend traditional views on reality and the core principles that dictate the cosmos. Pioneered by giants such as Planck, Bohr, and Heisenberg, the historical development of quantum theory has transformed contemporary physics and technology, heralding an epoch of quantum realities. At its crux lie concepts like superposition, wave-particle duality, and quantum entanglement — all pivotal in redefining our grasp on the minuscule universe while stretching scientific exploration's frontiers. The Schrödinger equation serves as a beacon within these theoretical underpinnings, revealing facets of a mysterious domain where particles simultaneously occupy manifold states. Interpretations such as many-worlds or Copenhagen further complicate quantum mechanics' mystique, provoking thoughts on existence's interwoven nature alongside boundless opportunities awaiting discovery within this esoteric realm.

Experimental Realizations

In the realm of deciphering quantum mechanics, its practical applications play a pivotal role, wherein theoretical notions undergo scrutiny in tangible setups. These experimental endeavors act as concrete affirmations of the esoteric concepts outlined within quantum theory, effectively narrowing down the divergence between abstract thought and tangible existence. Pioneering experiments such as the double-slit experiment along with Bell's theorem have ushered in compelling substantiation for puzzling phenomena like wave-particle duality and entanglement at a quantum level, thus shaking the foundational precepts of classical physics to their core. Through meticulous observation of particle behavior under meticulously controlled conditions, researchers are able to peel back layers surrounding quantum mechanics enigmas and delve into understanding reality's most basic essence. Not only do these empirical investigations enrich comprehension concerning quantum occurrences but they also lay groundwork for innovations across tech-domains including but not limited to quantum computing, encryption methodologies, and communicative infrastructures. This dynamic synergism betwixt theoretic conjectures and hands-on experimentation embodies the lively and continuously adaptive quest inherent within this intriguing segment of science.

Misconceptions and Clarifications

Often, misunderstandings about quantum mechanics stem from its unconventional aspects and how it diverges from traditional physics. A widespread but incorrect belief is that quantum entanglement permits particles to communicate instantly over any distance, whereas in truth, information cannot travel faster than light permits. Likewise, there exists a flawed perception regarding the observer effect; many believe observation directly affects a quantum experiment's outcome. Yet this effect subtly emphasizes measurement's importance in quantum setups rather than suggesting consciousness can alter results. It's vital for novices to disentangle these misconceptions to accurately comprehend the essential elements of quantum mechanics and admire its complex reality without being misled by false interpretations. By rectifying these misapprehensions and elucidating crucial notions clearly, beginners are equipped to tread through the intricacies of quantum mechanics with both certainty and insight.

XIII. Measurement Problem

The issue of Measurement in quantum physics presents a profound obstacle for grasping how quantum systems behave when observed. This critical concern probes into the essence of what constitutes reality and the observer's part in shaping outcomes. Upon observation, a quantum system transitions from being in a state of multiple possibilities to one definite state, sparking discussions over the operational processes and consequences for free will. The Problem of Measurement underscores the intrinsic ambiguity and chance-driven character of quantum physics, underscoring the necessity to revisit old concepts regarding causality and effect. For neophytes venturing into quantum mechanics, this notion acts as an intriguing threshold into the intricacies of a quantum domain where merely observing alters actuality fundamentally, questioning traditional perceptions about how the universe operates and its properties.

The Observer Effect

Venturing into the complex domain of quantum mechanics unveils the pivotal Observer Effect, a phenomenon emphasizing observation's significant impact on quantum systems, thus challenging conventional perceptions of objective reality. Quantum theory suggests that measurement itself modifies a particle's state, revealing a deep link between observer and observed. This mutual association indicates a deeper insight into how our observations shape reality's framework, merging the distinction between observer and observed. The reach of the Observer Effect surpasses physics' boundaries, engaging in philosophical debates about consciousness, autonomy, and existence essence. Welcoming this quantum principle provides novices with an eye-opening notion that underlines quantum mechanics' elaborate interconnections and its profound effects on understanding cosmosity.

Collapse of the Wave Function

When exploring the basics of quantum mechanics for novices, it becomes crucial to shed light on 'Collapse of the Wave Function' as a key idea needing clarity. The process where the wave function, which encapsulates potential states of a quantum system, condenses into a singular state upon measurement is fraught with implications both philosophical and functional. This instance of collapse disrupts conventional views on predictability and injects unpredictability into the domain of quantum physics. Comprehending this event is vital in understanding how measurements impact quantum entities and the significant role uncertainty occupies within quantum mechanics principles. By demystifying this notion effectively, those new to the subject can start to comprehend the fundamental rules that dictate the behavior in the quantum universe and grasp its intricacies at decoding such sophisticated realms.

Interpretations and Controversies

When exploring the complex world of quantum mechanics, one stumbles upon various interpretations and debates that disrupt conventional notions of existence and the cosmos. At the heart of these arguments lie theories such as the Copenhagen and Everett interpretations, presenting divergent takes on quantum phenomena's probabilistic characteristics. The Copenhagen Interpretation suggests probability-driven outcomes in quantum events, in sharp contrast to the Everett Interpretation, which images a universe where every potential outcome coexists at once. These opposing viewpoints ignite fervent dialogues among thinkers and scientists, evoking deep reflections on fate, autonomy, and what it means to be. The doubts and intricacies entwined within these perspectives highlight the profoundness of quantum mechanics, urging us to question our basic beliefs about reality.

XIV. Copenhagen Interpretation

In the framework of "Deciphering Quantum Mechanics: Key Concepts Explained for Beginners," devised by notables such as Niels Bohr and Werner Heisenberg, the Copenhagen Interpretation emerges as a groundbreaking pillar in unlocking the obscure universe of quantum phenomena. It theorizes that until an observation occurs, quantum entities linger in various states of superposition, disrupting traditional deterministic perspectives on reality and paving the way for probabilistic results. This interpretation acts as a crucial portal for neophytes striving to comprehend the intrinsic uncertainty prevalent within quantum systems. By casting light on how observations mold both behavior and characteristics of particles, it provides rudimentary insights into the fluid and intertwined essence of existence at the microscopic scale. The philosophy enshrined in this interpretation is conductive to reflection and analytic thought processes—indispensable tools for beginners threading through quantum mechanics' complexities. For those newly initiated into dissecting quantum conundrums, it lays down an essential blueprint aiding their early ventures and formulations.

Overview of the Interpretation

Modern physics is fundamentally revolutionized by quantum mechanics, a pivotal force that challenges traditional beliefs and unveils the secrets of the microscopic world. This transformative theory emerged from the seminal work of pioneers such as Niels Bohr, Max Planck, and Werner Heisenberg, reshaping our understanding with critical notions. Concepts like entanglement and superposition reveal a universe deeply connected at its core, encouraging newcomers to engage with the peculiarities of wave-particle duality and indeterminacy. As novices traverse through this theoretical domain, they dissect various interpretations including those proposed in Copenhagen or by Everett, confronting profound realizations about stochastic results and ambiguity. In grasping these essential ideas along with their historical roots, beginners set off on an expedition to demystify quantum mechanics' enigmatic wonders armed with eagerness and inquiry.

Philosophical Underpinnings

The exploration of quantum mechanics offers deep insights into how this groundbreaking theory transforms our view of reality and existence. The entwined characteristics of quantum entities throw traditional deterministic perspectives into question, fostering philosophical musings on consciousness and the concept of free will. Phenomena such as the observer effect and quantum entanglement underscore a vital connection between observation and reality, provoking deliberation regarding human consciousness's impact on shaping the cosmos. By accepting uncertainty and interconnection, people can adeptly maneuver through the intricacies of quantum phenomena, adopting more adaptable strategies for making decisions while gaining an enriched comprehension of reality's fluid nature. Quantum mechanics reshapes not only our technological framework but also encourages a reevaluation of our core beliefs about the universe's essence and our role amidst it.

Criticisms and Alternatives

Delving into the Disputes and Substitutes related to quantum mechanics, recognition of the debate-laden interpretations and the obstacles they introduce to our grasp on the quantum universe is paramount. Detractors frequently highlight the enigmas tied to effects caused by observation and the counterintuitive phenomenon known as quantum entanglement. These objections scrutinize the built-in unpredictability and subjective impacts in quantum occurrences, casting uncertainty over quantum mechanics' aptitude as an exhaustive framework. On flip side, alternative perspectives advocate for more investigation and fine-tuning rather than complete negation. Through contemplating divergent constructs or explanations like pilot-wave theory or theories with unseen variables, scholars strive to tackle traditional quantum mechanics' shortcomings and ambiguities. This continuous exchange between critiques and variant methodologies illuminates on the vibrant, shifting essence of quantum paradigm; it beckons an intensified inquiry into its intricacies by novices eager to decode its base principles.

XV. Many-Worlds Interpretation

In the 1950s, Hugh Everett introduced the Many-Worlds Interpretation (MWI) within quantum mechanics, suggesting the fascinating concept that multiple parallel universes emerge with each quantum occurrence, thus presenting a novel viewpoint on reality's essence. According to this theory, every conceivable result of a quantum experiment actually happens, giving birth to an extensive array of simultaneous universes for each potential outcome, thereby questioning established notions of determinism and cause-and-effect relationships. MWI proposes a captivating model that accommodates quantum phenomena's inherent unpredictability and provides expansive insights into the principles of quantum superposition and entanglement. Pondering over MWI's propositions enables novices to understand how everything is interlinked and appreciates the evolving aspects of quantum systems, leading them towards an intricate journey in understanding the complexities of the quantum world beyond traditional limits.

Explanation of Everett's Theory

The Many-Worlds Interpretation, also identified as Everett's Theory, captivates both philosophers and physicists with its radical reconceptualization of quantum mechanics. Formulated by Hugh Everett during the 1950s' culmination, this notion introduces the idea that each quantum assessment bifurcates the cosmos into diverse parallel existences, portraying alternative potential outcomes. This hypothesis upends conventional perspectives on reality and observation by implying that every conceivable conclusion transpires concurrently, wherein each divergent universe maintains a separate yet linked existence. For neophytes venturing into quantum mechanics' domain, assimilating Everett's Theory could be simultaneously bewildering and illuminating—unveiling the profound depths and enigmas encompassed within the quantum dimension. Venturing through this unconventional interpretation aids initiates in comprehending the significant repercussions of quantum physics upon our delineation of reality and cosmic understanding.

Implications for Reality

Challenging the traditional viewpoints, the implications within quantum mechanics for reality reveal a complex web of interconnections and fluid characteristics among quantum particles. The phenomena of superposition and entanglement in quantum theory contradict our usual ideas about cause and effect, leading to questions about our deterministic frameworks. Experiments like the double-slit test alongside Heisenberg's Uncertainty Principle highlight the intricacies within quantum mechanics, encouraging a fresh look at how we view existence and reality itself. As progress in technologies based on quantum principles such as secure communications and computing progresses, it fuels deeper philosophical debates around concepts like free will vs determinism and the fundamental essence of the cosmos. By accepting ambiguity and unpredictability as part of scientific inquiry, new avenues emerge that encourage adaptable methods to confront the enigmatic aspects of quantum mechanics—ultimately transforming our collective comprehension concerning both reality and the broader cosmos.

Debate and Acceptance

Within the domain of quantum mechanics, a significant area of disagreement is centered on the debate and subsequent acknowledgment of its intricate concepts. Groundbreaking figures such as Niels Bohr, Max Planck, and Werner Heisenberg established the foundation for this transformative discipline, yet further interpretations and uses have ignited academic debates and philosophical discussions. Ideas like wave-particle duality, superposition, and quantum entanglement defy traditional physics principles and necessitate thorough scrutiny. The probabilistic Copenhagen Interpretation encounters challenges from competing theories like the many-worlds interpretation. Despite these scholarly conversations about quantum mechanics, its practical applications in fields like cryptography and quantum computing are undeniable influences on the trajectory of scientific progress and innovation. It's essential to engage with these discussions to propel our comprehension of the cosmos forward and tap into quantum phenomena's capabilities for societal advancement.

XVI. Decoherence Theory

Theory of Decoherence, an imperative notion within the realm of quantum mechanics, is pivotal for grasping how we move from quantum to classical realities. This theory explicates the unavoidable engagement between a quantum entity and its surrounding environment, causing what seems like a collapse of the entity into a state that's classical. It addresses how quantum coherence is lost and how classical actions appear without direct measurement involved, illuminating on where the tiny-scale quantum domain ends and our large-scale observable world begins. Following how states in quantum mature along with their entanglement across surroundings, decoherence theory serves as a skeleton key in understanding how behaviors typically seen as classic originate from systems fundamentally rooted in quantum. Incorporation of this theoretical framework into more extensive discussions around the mechanics of quanta aids novices in gaining essential insights regarding interactions between quantum entities and their environments, thereby setting groundwork for further delving into complexities enveloping our understanding of realms shaped by quanta.

Role in Quantum Mechanics

In the domain of quantum mechanics, the role of the observer is crucial in molding reality and affecting outcomes within quantum systems. This idea emphasizes a deep link between consciousness and how quantum entities behave, undermining views that predict universe behavior with certainty. Experiments in the realm of quantum have shown how human awareness affects particle states, indicating an indirect connection between those observing and what is observed. The effect of observing underscores a complex relationship between one's perception and actual physical existence, leading to philosophical musings about life's essence and autonomy. By recognizing the influence observers have on phenomena at the quantum level, we pave pathways for further exploration into how consciousness may shape universal fabric while encouraging novices to investigate more about quantum mechanics' enigmatic nature along with its philosophical foundations.

Explanation of Classical Transition

Delving into the classical-to-quantum transition in physics, we witness a pivotal shift from traditional classical theories towards the principles of quantum mechanics, signifying a deep transformation in how we perceive the universe. This shift is highlighted by the surfacing of quantum behaviors such as entanglement and superposition, which question long-held beliefs about causality and determinism. As it investigates particle behavior at both atomic and subatomic scales, quantum mechanics reveals that the old classical models fall short of accounting for complex interactions and unforeseeable outcomes characteristic of quantum domains. In this new quantum era, particles are discovered to coexist in numerous states at once—a phenomenon referred to as superposition—and to become intertwined in such a way that changing one particle's properties can instantaneously affect another's across vast distances. Moving away from classic physics reveals deeper layers and connections within the microscopic structure of reality, altering our understanding fundamentally while beckoning novices to explore these tantalizingly intricate nuances inherent within quantum mechanics.

Resolving Measurement Problems

In the domain of quantum mechanics, addressing issues related to measurements takes center stage as a crucial obstacle demanding innovative thinking and resourceful solutions. Basic tenets like entanglement and quantum superposition introduce a degree of unpredictability in the outcomes of measurements, rendering old-school measurement techniques inadequate. Delving into the complexities tied to particle-wave duality and how observation impacts results allows scientists to wade through the complications associated with quantifying phenomena within quantum systems. Ideas such as Heisenberg's Uncertainty Principle illuminate the bounds on accuracy within quantum dimensions, underscoring the urgency for overhauling traditional approaches towards gauging phenomena. The realm of quantum mechanics furnishes practical uses in realms including but not limited to cryptography and computational prowess using quanta, compelling researchers to reformulate ideas on capturing these elusive potentials entirely. Quantum mechanics disrupts conventional perspectives regarding reality via marvels like superposition alongside entanglement. Such revelations find functional relevance across domains spanning from computing driven by quanta to biological spheres wherein peculiarities at a subatomic level optimize processes involving energy conveyance alongside chemical changes. The confluence between neuroscience and principles governing quanta arouses debates surrounding consciousness plus potential influences exerted by quantitative occurrences upon neural activities—probing further into these junctures promises revolutionary strides across healthcare sectors, biotechnological advancements alongside ecological conservation efforts; implications borne by minute effects might alter comprehending cognitive functionalities.

XVII. Quantum Field Theory

As a crucial subset of quantum mechanics, Quantum Field Theory explores the complex interactions among particles and fields, transforming our perception of the cosmos. This theory expands upon quantum mechanics by treating fields as primary elements, thereby disclosing the essence of particles and forces within nature. Pioneered by figures such as Dirac and Feynman, it sheds light on how field excitations give birth to particles and delineate their interplay, offering an insightful glimpse into the minuscule realm. Serving as a cornerstone for contemporary physics, this theoretical construct not only propels advancements in technology like those observed in semiconductors but also stimulates innovation across numerous domains. By familiarizing themselves with Quantum Field Theory's basics, novices can adeptly traverse through quantum complexities from an enriched vantage point that enhances overall appreciation for reality's intricacy and its boundless opportunities.

Introduction to Fields in Quantum Mechanics

Laying the foundation for comprehending the minutely interconnected cosmos, an initiation into quantum mechanics fields defies traditional perceptions of existence. Through these fields, particles transmit and interact, molding the actions of quantum beings. Quantum field theory is a realm pioneered by notable scientists such as Dirac and Feynman; it penetrates deep into the rules that orchestrate these natural forces and interactions. Diving into the complexities of quantum mechanics' fields permits novices to understand core ideas molding contemporary physics alongside technological advancements. These areas act as a prism revealing insights like particle weaving together (entanglement), layers upon state (superposition), and how particles mingle, thereby enlightening one on quantum mechanics' arcane domain. Comprehending field roles stands pivotal for decoding quantum incidents' intricacies along with their significant outcomes in technical realms and scientific exploration.

Unification of Forces

In the endeavor to decode the enigmas of cosmos, merging forces emerges as a pinnacle pursuit within quantum mechanics' scientific exploration domain. As it ventures into the microscopic realm, quantum theory aims to amalgamate nature's fundamental forces—electromagnetic, weak nuclear, strong nuclear, and gravitational—into a unified schema. This quest for amalgamation not merely strives to blend apparently incongruent occurrences but likewise heralds an enriched comprehension of reality's foundational weave. By scrutinizing how these forces interact at a quantum scale, scholars aspire to unearth possibilities for groundbreaking progress in both technological realms and theoretical physics principles. The path toward merging epitomizes the core of unraveling quantum mechanics complexities while presenting novices with insights into the sophisticated intricacies that regulate cosmos operation and ignite scientific inquiry.

Standard Model of Particle Physics

The Particle Physics Standard Model is a critical framework for comprehending the varied interactions among particles and forces across the cosmos, encapsulating essential principles that outline the subatomic realm. Crafted through detailed experimentation and theoretical investigation, this model symbolizes an elaborate union of significant scientific insights from notable individuals like Max Planck, Niels Bohr, and Werner Heisenberg. At its heart, the Standard Model illuminates complex relations between elementary particles and primal forces, drawing a unified story about matter's foundational elements and their governing laws. Incorporating entities such as quarks, leptons, and bosons in one cohesive explanation, this model acts as a guiding light within quantum mechanics' domain, providing an organized and exhaustive summary crucial for novices to understand the captivating intricacies of the subatomic world.

XVIII. Quantum Gravity and the Search for Unification

Embarking on a journey to integrate quantum mechanics' domain, the probe into quantum gravity emerges as a crucial pursuit, endeavoring to reconcile the discrepancies between general relativity and quantum mechanics. The quest for Unification and Quantum Gravity signifies an intense effort to meld these two critical cornerstones of physics together, shedding light on cosmic operations at both extensive and minute levels. Venturing towards a comprehensive theory that includes gravity under quantum mechanics' umbrella necessitates complex theoretical constructs and mathematical models, overturning traditional views and extending the frontiers of scientific exploration. Through examining the complexities inherent in quantum gravity, scholars aim to decode cosmic enigmas, fostering a richer comprehension of reality's unified essence and setting the stage for novel breakthroughs in theoretical physics plus cosmology.

Challenges in Unifying Gravity with Quantum Mechanics

Merging gravity with quantum mechanics stands as a primary obstacle in contemporary physics, bewildering researchers for numerous years. The conflict stems from the distinct differences in scale at which both theories operate—cosmic expanses are the domain of gravity, while the infinitesimal world is under quantum mechanics' sway. Referred to as quantum gravity, this endeavor to unify these critical components of physical theory has yet to find resolution because of their complex mathematical structures and our current experimental capabilities' constraints. This absence of a cohesive framework hampers advancements in grasping spacetime's behavior at subatomic levels and creates substantial hurdles in integrating nature's essential forces. The sophisticated relationship between gravitational forces and quantum phenomena continues to captivate and confound academia, highlighting the daunting task of unraveling the mysteries that pervade the quantum cosmos.

Approaches to Quantum Gravity

Various theoretical constructs aim to bridge the gap between quantum mechanics and general relativity, addressing the fundamental essence of spacetime. A key theory in this quest is Loop Quantum Gravity, which proposes that spacetime's geometry is quantized into discrete units. This hypothesis endeavors to tackle classical general relativity's singularity issue while offering perspectives on the spatial and temporal properties at microscopic scales. Conversely, String Theory suggests that universe's most basic constituents are not zero-dimensional points but oscillating strands. These filaments generate diverse particles based on their oscillation patterns. Although both theories offer captivating insights into gravity's quantum aspects, they encounter obstacles regarding empirical confirmation and integration with additional primary forces. Delving into these theories serves as an introduction for newcomers to the intricate world of quantum gravity, illuminating ongoing research efforts that redefine our cosmic comprehension.

Significance for Cosmology

Quantum mechanics holds a critical relevance for cosmology, illuminating the core essence of the universe and its chronological development. Through investigation into occurrences such as entanglement and cosmic microwave background radiation, quantum mechanics has ushered in a paradigm shift in cosmological studies, offering an alternative perspective to unravel the cosmos's intricacies. It contests conventional notions regarding time, space, and mutual connectivity, enlightening on how interconnected the cosmos truly is. The examination of quantum entanglement alongside the impact of quantum phenomena on cosmic formations intimates at a profound connection between quantum happenings and the emergence of vast universal structures. This confluence of quantum mechanics with cosmology not merely deepens our comprehension of the cosmos but also drives forward scientific exploration into uncharted territories within theoretical physics thereby molding our understanding about reality and existence.

XIX. Philosophical Implications

Delving into the complexities of quantum mechanics surpasses the conventional boundaries of physics, touching upon profound philosophical questions inherent within this complex domain. This field contests basic assumptions regarding determinism, paving paths for debate over reality's essence and consciousness itself. Ideas such as quantum entanglement and the observer effect posit a profound unity among all constituents, evoking deep reflections on existence, time-space continuum, and human life significance. In accepting the uncertainties that come with quantum phenomena, people are encouraged to approach decision-making and forming relationships with greater openness. Integrating quantum principles into daily living can guide individuals towards an enhanced sense of connectedness, highlighting personal evolution's fluidity alongside the importance of awareness and inventiveness in addressing both the intricacies of quantum mechanics and wider philosophical quests.

Reality and Objectivity

Within the universe of quantum mechanics, conventional conceptions of objectivity and reality are transformed, presenting challenges to established norms. Phenomena such as entanglement and superposition within this realm oppose the ideas of determinism and classical causality, leading to questions regarding the solidity and impartiality of what we perceive as real. According to the fluid connection between quantum elements necessitates a reconsideration of our grasp on existence, diminishing the distinction between those observing and that which is observed. The action of measuring in this microscopic domain shifts outcomes, underscoring how subjective perspectives affect what is seen as objective truth. This complex interaction among observers and phenomena underlines a critical interplay between subjectively held beliefs and universally accepted realities in quantum mechanics. For beginners embarking on deciphering its cryptic rules, understanding these effects on notions like reality becomes crucial for demystifying this dimension's secrets.

Determinism and Free Will

Exploring the complex domain of quantum mechanics unveils deep inquiries about the juxtaposition of determinism versus autonomy. The conventional deterministic perspective, which suggests events are predestined, is at odds with the probabilistic essence of quantum mechanics, where outcomes remain indeterminate until they're measured—this interrogation challenges our grasp on actuality. The Copenhagen Interpretation theorizes that measuring action alters a system's state, integrating a factor of volition and unpredictability that smudges distinctions between determinism and autonomy. Quantum oddities such as entanglement and superposition add layers to this contrast, hinting at a cosmos steered by likelihoods over fixed outcomes. As novices navigate through quantum mechanics' intricacies, discussions surrounding determinism versus free will pave pathways for philosophical speculation and reassessment of our cosmic significance.

Quantum Mechanics and Consciousness

Delving into the fascinating bond between consciousness and quantum mechanics reveals a complex interplay that disrupts our usual views on reality and awareness. Phenomena within quantum mechanics, like entanglement and superposition, suggest a deeper linkage across the cosmos that reaches into consciousness itself. The significance of an observer in quantum scenarios showcases how human perception influences outcomes, igniting debates over determinism versus free will. This complex connection highlights possibilities for gaining insights into human consciousness and the webbed nature of existence by examining their convergence. Through such exploration, we breach established limits and launch onto an intriguing path that reshapes our grasp of consciousness intertwined with quantum mechanics' core aspects.

XX. Quantum Mechanics in Popular Culture

Quantum mechanics, with its deep impact, has woven into the fabric of popular culture, leaving a mark on artistic expression through varied channels. Movies such as "Inception," which explores the convoluted realms of dream manipulation and reality's essence to TV series like "The Big Bang Theory" that lightheartedly weave quantum theory into daily vignettes, showcase how quantum mechanics is repeatedly embroidered in entertainment tapestries. Literature too isn't untouched; sci-fi works by luminaries like Isaac Asimov and Philip K. Dick delve into quantum-laden themes including alternate realities and chrononautics, entrancing audiences with plots that twist perception. Video gaming offerings such as "Quantum Break" and "BioShock Infinite" cleverly integrate quantum tenets into their core play strategies, gifting gamers experiences where fiction marriers reality in unforeseen ways. These cultural artifacts do more than amuse; they double as didactic instruments, ushering complex quantum ideas onto a broad stage in an engaging and digestible format. The enmeshment of quantum mechanics within pop culture signals its escalating role in molding our grasp of existence while fueling inventive ventures that leap over established conceptual confines.

Misconceptions and Exaggerations

Confusions and overstatements regarding quantum mechanics often obscure the core principles underlying this complex area of physics. Popular confusions, like believing that quantum mechanics is restricted to tiny scales, miss understanding the wider effects of quantum phenomena in the larger world. Overstatements, such as the claim that quantum mechanics allows for immediate message transmission over long distances, simplify too much the intricacies of quantum entanglement and its actual constraints. These confusions and overstatements obstruct a newcomer's comprehension of essential quantum concepts like superposition and wave-particle duality, which are crucial for truly appreciating what defines the realm of quanta. By clearing up these confusions and accurately defining both what is within reach and outcomes from mastering quirks made by atomic science parts; novices might proceed through this bewildering territory with exactitude plus nuance—establishing an unwavering groundworks during venturing further into this enchanting domain.

Influence on Literature and Film

Delving into the impact of quantum mechanics across literature and film reveals its broad swath through various artistic fields. By introducing convoluted notions such as quantum entanglement, it emphasizes the universe's intrinsic connectivity, fueling the creative fires of both authors and directors. Narratives in books frequently interlace with elements from quantum theories, upending conventional views on reality and awareness. In a parallel manner, films weave in aspects of quantum thinking to probe into life's existential enigmas, softening distinctions between scientific inquiry and philosophical speculation. The realm of storytelling has been enriched by quantum mechanics, nudging human creativity to explore beyond known frontiers while engaging audiences in contemplative journeys framed by spellbinding tales and visuals. This melding together of sciences with the arts for probing beneath quantum phenomena layers subtlety and fascination over narrative vistas; this proffers spectators an unusual perspective for untangling universal complexity.

Public Understanding and Interest

The role of public comprehension and enthusiasm for quantum mechanics is vital in molding societal perspectives toward this intricate subject area. By clarifying core notions such as superposition and quantum entanglement with easy-to-understand explanations and comparable instances, people from different scientific understandings can grasp and value the key principles of quantum mechanics. Efforts to boost public interest, including instructional seminars and engaging digital content presentations, are essential in sparking curiosity and advancing knowledge about quantum phenomena across a variety of groups. When individuals take an active part in unraveling the enigmas of quantum science, it triggers a domino effect that elevates general scientific awareness and spawns significant conversations regarding both the practical uses and deep-seated philosophical questions raised by quantum mechanics. In essence, fostering public insight into and excitement about quantum mechanics not only augments our shared intelligence but also lays down tracks for groundbreaking developments in science technology which could transfigure our global outlook moving forward.

XXI. Educational Approaches to Quantum Mechanics

Approaches to teaching quantum mechanics are crucial for rendering this intricate topic understandable for learners, especially those new to the field. By organizing learning resources to sequentially introduce essential notions such as superposition, entanglement in quantum physics, and the dual nature of particles and waves in an engaging and lucid way, teachers can unravel the complexity of quantum mechanics for beginners. Leveraging examples and visual aids proves instrumental in elucidating these core ideas, making a connection between abstract theories and their real-world applications. The inclusion of theoretical frameworks like Schrödinger's formula along with popular interpretations such as Copenhagen's and Everett's provides a basic comprehension without deluging learners with complex mathematical details. Through the use of easy-to-understand language combined with trustworthy academic references, educational narratives manage to navigate novices across the convoluted landscape of quantum mechanics whilst igniting enthusiasm and analytical thought regarding its subsequent role within science and technological innovation sectors.

Teaching Complex Concepts

For beginners to grasp the intricate subject of quantum mechanics, a thoughtful educational strategy is vital for effective engagement. Through employing various teaching techniques including illustrations, dynamic activities, and basic comparisons, the chasm between esoteric ideas and perceivable comprehension can be narrowed. Embedding examples from everyday life alongside practical uses of quantum theories allows teachers to render the topic more approachable and captivating for pupils. Promoting student involvement, analytical reasoning, and tangible experimentation enhances mastery and recall of crucial concepts. Additionally, establishing an encouraging learning atmosphere that invites inquiries and values differing viewpoints improves the quality of education students receive. Adapting pedagogical tactics to suit individual learning preferences and knowledge backgrounds enables educators to clarify complicated subjects like quantum mechanics, granting newcomers confidence in exploring quantum intricacies.

Use of Simulations and Visualizations

In making the intricate ideas of quantum mechanics understandable, especially for novices endeavoring to comprehend its basic tenets, simulations and visualizations are paramount. Interactive explorations through simulations enable learners to experience abstract principles such as superposition and quantum entanglement in a form that is both visually appealing and more graspable. The mystery surrounding quantum entities' behavior, like wave-particle duality, becomes less opaque with the help of visualizations offering solid illustrations that promote comprehension. These instruments serve as a pivotal bridge linking theoretical notions to their practical implications, thus simplifying the conceptualization of quantum phenomena for individuals. By embedding these tools within teaching resources – be it digital platforms or hands-on workshops – beginners gain enhanced insights into the complex realm of quantum mechanics which lays down a foundation for enriched learning encounters and an expanded understanding of this captivating discipline.

Encouraging Intuitive Understanding

Fostering an intuitive grasp of quantum mechanics is critically important for novices exploring the intricacies of this challenging area. By nurturing a perspective that accepts unpredictability and interconnection, people can achieve an instinctive understanding of quantum ideas. Careful consideration and reflection on various possibilities and likelihoods can improve one's ability to make decisions and enrich their perception of the quantum world. This strategy highlights the fluid character of quantum occurrences and stresses the significance of integrating personal advancement with quantum theories. Through incorporating mindfulness techniques into everyday activities and aiming for objectives inspired by quantum creativity, individuals are able to develop an instinctual and comprehensive tactic toward solving problems. Promoting intuitive comprehension not only boosts understanding of quantum mechanics but also fosters a mentality characterized by receptivity and versatility in dealing with the complexities inherent within this intriguing domain.

XXII. Quantum Mechanics and Metaphysics

Delving into the intricate relationship between Quantum Mechanics and Metaphysics unveils a captivating examination of existence and reality's essence. The axioms of quantum mechanics disrupt entrenched beliefs about cause-effect relations and predictability, probing instead into a universe characterized by probabilities and interconnections. Phenomena such as quantum superposition and entanglement upend our conventional views on space-time dimensions, along with how the observer impacts the fabric of reality. Quantum mechanics' revelations reach far outside mere physical confines, stirring reflections on consciousness, autonomy, and the universe's intrinsic qualities. For novices unraveling quantum mechanics' core tenets, they are invited to reflect on its metaphysical ramifications—discovering an elaborate network of unpredictability and connection that surpasses ordinary scientific discourse borders. This journey marks an entry point for achieving profound comprehension concerning the cosmos and humanity's role therein; it encourages speculation about how quantum realities interlace with metaphysical domains.

Interplay Between Physics and Philosophy

Exploring the overlap of physics with philosophy, quantum mechanics discloses a deep link between explorations in science and existential thought. Unveiling complexities within quantum realities provokes a questioning of traditional philosophical beliefs, necessitating a reassessment of reality's core essence and our role therein. Phenomena such as superposition and quantum entanglement merge philosophical dialogues with physical events, stirring debates on topics like free will, determinism, and how observers influence the unfolding of reality. By exposing the universe's fluidity and interconnected nature through quantum mechanics, it encourages conversations about perception, consciousness, and existence's structure. The merger between philosophy and physics not just enhances comprehension of the quantum domain but also delivers substantial reflections on knowledge, reality, or human insight — bridging scientific examinations closely with meditative pondering.

Questions About the Nature of Existence

Inquiries regarding the essence of being come to light amidst the intricacies of quantum mechanics, putting conventional perceptions of reality under scrutiny and sparking philosophical musings on the universe's unified nature. Ideas such as superposition and quantum entanglement reveal a more intricate foundation beneath existence, prompting profound contemplations about free will and consciousness. Quantum phenomena's impact on human perception and choice-making appears significant, hinting at a fluid and unpredictable essence of reality. By accepting quantum mechanics' vagueness and reconsidering fixed models, people can tread through uncertainty with an adaptable mindset open to comprehending beingness and the cosmic network it weaves. Diving into these inquiries not only transforms our grasp of what is real but also provokes deeper reflection upon our own awareness within this enigmatic quantum framework.

Impact on Theological and Metaphysical Thought

The influence of quantum mechanics on theological and metaphysical reasoning is significant, undermining traditional views on existence and reality. Phenomena such as superposition and entanglement in the quantum realm introduce a notion of unpredictability and interconnectedness that aligns with wider philosophical debates. The idea that particles can display behavior akin to waves and be in several states at once prompts deliberation over consciousness's nature and how observation might mold reality. Principles like the Uncertainty Principle alongside the observer effect imply that outcomes in quantum systems are impacted by human perception, obscuring distinctions between the observed and the observer. This merging of quantum concepts with metaphysical theories ignites discourse surrounding determinism, free will, and universe's essence, encouraging reflections on how scientific exploration interacts with existential pondering. Venturing into these dialogues paves fresh pathways for metaphysical and theological discussion, unsettling established convictions while inviting reconsideration of core tenets under the illumination provided by quantum discoveries.

XXIII. Quantum Mechanics in Biology

In the domain of life sciences, the fusion with quantum mechanics reveals an intriguing dynamic between the realm of the minutely small and living entities, providing fresh outlooks on the inner workings of life. Viewed through quantum mechanics' perspective, occurrences such as quantum tunneling and coherence illuminate how photosynthesis and enzyme activities achieve their efficiency and complexity. Quantum entanglement could influence the complex web within biological systems, affecting everything from how genes interact to how neural networks function. Grasping how quantum effects might shape biological systems paves pathways for investigating ways in which quantum concepts can deepen our understanding of existence at both a molecular scale and more expansively. Venturing into this confluence of quantum mechanics with biology propels us towards decoding existential enigmas via a quantal vantage point, igniting groundbreaking revelations while unsettling established views in biosciences.

Quantum Effects in Biological Systems

The intriguing overlap of quantum mechanics with the complexity of life forms highlights an area where biology and quantum principles intersect. The exploration into how quantum fundamentals can be applied to biological activities is drawing more examination due to its prospective consequences. Investigating microscopic occurrences such as superposition and quantum entanglement, scholars are revealing potential ways these effects might impact functions within organisms. For example, it has been proposed that enzymatic actions could involve quantum tunneling, presenting a challenge to the traditional understanding of biochemical operations. Additionally, investigating coherence in living systems prompts inquiries into how quantum events might influence processes like animal magnetoreception and photosynthesis. Gaining insights into the role of quantum phenomena in life mechanisms not only provides a fresh outlook on biological workings but also suggests a deeper connection across the universe at previously uncharted magnitudes, enriching our grasp on both living entities and the realm of quantum mechanics.

Quantum Biology Research

Research in quantum biology is an intriguing overlap of quantum mechanics with life sciences, injecting fresh viewpoints on the workings of life. Investigating molecular-level phenomena such as superposition and entanglement, scientists seek to demystify biological functions with a level of precision and insight never seen before. Quantum mechanics brings to the table ideas of connectivity and unpredictability that could shine light on various processes like photosynthesis, reactions catalyzed by enzymes, or even activities within neurons. The infusion of quantum principles into biology paves the way for exciting revelations about how organisms operate at their core levels, hinting at revolutionary advancements in areas such as medicine and biotechnology. As research in quantum biology progresses, it opens up thrilling possibilities for cross-disciplinary cooperation and breakthroughs, narrowing the divide between theories of quantum physics and biological realities to expose the complex beauty underlying living systems.

Implications for the Study of Life

For the exploration of life, quantum mechanics posits a profound shift, presenting an alternative perspective for deciphering biological systems' intricacies. Phenomena such as entanglement and superposition in quantum mechanics hint at a degree of interconnection and non-locality, potentially impacting biological operations. The phenomenon where particles can occupy various states at once challenges our traditional deterministic views on biology's workings, suggesting that quantum influences might play a role in how molecules interact and cells function. This reinterpretation of existence through the quantum spectrum could shed light on previously inexplicable aspects of biological system behaviors, deepening our comprehension of life's mechanisms. Investigating the overlap between life sciences and quantum mechanics might reveal unprecedented biological functioning principles, setting the stage for revolutionary advancements in areas like neuroscience and molecular biology.

XXIV. Quantum Mechanics and Chemistry

When examining the intersection of Quantum Mechanics and Chemistry, it is crucial to dive into the core principles that dictate how atoms and molecules behave for a deep grasp of chemical processes at a quantum scale. The advent of quantum mechanics has transformed chemistry by shedding light on how atoms bond, the energy levels within molecular orbitals, and how chemical reactions unfold. Key notions such as superposition and quantum entanglement are critical in clarifying particle behavior at a tiny scale, impacting both stability and reactivity in chemicals. The concept of wave-particle duality provides an insightful view into matter and light's dual characteristics, enhancing our understanding of molecular configurations and spectroscopic methods. By exploring the complex link between quantum mechanics and chemistry deeply, we unravel the dense network controlling matter's basic constituents alongside the elaborate sequences of chemical alterations; thereby highlighting quantum mechanics' substantial influence over chemistry's domain.

Chemical Bonding and Reactions

Quantum mechanics serves as the theoretical bedrock for demystifying behaviors of matter at a molecular scale, emphasizing its paramount role in chemical bonds and reactions. Central to understanding chemical bonding lies the mechanisms of atomic interactions and electron sharing, depicted through quantum mechanical concepts like entanglement and superposition principles. Quantum theory shines a light on electrons' wave properties and their distribution across molecules, unraveling the complexity behind molecular configurations and reactivity traits. The pivotal Schrödinger equation in quantum physics propels our grasp on how electrons navigate chemical landscapes, illuminating pathways of bond creation and dissolution. Viewing chemical transactions through the prism of quantum phenomena allows us to perceive them as vibrant processes steered by electron probabilities and energetic states, offering an intricate yet profound vista onto the realm of molecules that fuels myriad technological advancements in chemistry fields.

Quantum Chemistry and Computational Methods

Within the sphere of quantum mechanics, computational techniques and quantum chemistry take on an essential function in revealing the characteristics and actions of molecules as well as atoms at a quantum scale. This field employs algorithms along with computational gadgets to model and scrutinize complex systems governed by quantum rules, shedding light on molecular configurations, chemical bonds, and pathways of reactions with accuracy that is unmatched. Leveraging the capabilities of computers based on quantum theories alongside intricate algorithms allows scientists specializing in this branch to decode complex occurrences within molecules which were once beyond reach via traditional computing methods. These sophisticated approaches do more than just amplify our grasp over chemical interactions; they also lay groundwork for breakthroughs in areas like material production, pharmaceutical development, and catalyst design. With the collaboration between computational strategies underpinned by principles from quantum physics constantly pushing scientific limits further outwards creates an optimistic prospect for exploration as well dynamic progressions across various sectors of chemistry research.

Advances in Material Science

The principles of quantum mechanics have deeply influenced advancements in material science, leading to groundbreaking innovations across multiple sectors. Phenomena such as entanglement and superposition have cleared the path for producing materials that boast unprecedented qualities and functions. By exploiting these quantum notions, experts have managed to create materials that exhibit superior flexibility, durability, and electrical conductance, transforming industries from health care to consumer electronics. The synergy between material science and quantum mechanics highlights an endless horizon for inventing state-of-the-art materials capable of meeting the dynamic demands of society while expanding the limits of known science. With a profound comprehension of quantum theories, scientists specializing in materials are spearheading initiatives toward an era where materials actively facilitate technological growth and societal development rather than merely existing as inert elements. Delving into the application of quantum mechanics within material science reveals opportunities that fuse imaginative speculation with tangible scientific achievement.

XXV. Quantum Mechanics and Astrophysics

Within the domains of quantum mechanics and celestial physics, a complex interconnection unravels, exposing the deep links between the minuscule and vast realms. Quantum mechanics provides a distinctive perspective for probing the universe's enigmas through its concepts of superposition and entanglement. By investigating wave-particle duality and the sophisticated interactions among quantum elements, we access an enhanced comprehension of celestial phenomena and galactic formations. The implementation of quantum principles in celestial studies reveals reality's essential structure, influencing our insights into space objects and cosmic occurrences. Bridging from particles to galaxies, this fusion between quantum science and astronomical studies urges us to reflect on both unity and diversity within the universe, driving us toward new horizons in knowledge discovery.

Quantum Phenomena in Space

Within the immense void of space, the phenomena of quantum mechanics unveil a deep linkage across the universe, pushing past our usual conceptions of existence. Quantum entanglement demonstrates how particles can stay linked regardless of their separation in space, putting into question our ideas on spatial limitations. This occurrence points to a universal network of interactions which bypass traditional physics rules and indicate an underlying unity within the cosmos's structure. The repercussions from these insights do more than pique scientific interest; they signal a profound alteration in perceiving space's very essence. Delving into quantum occurrences in the celestial expanse reveals a complex interplay of intertwined particles that traverse emptiness, emphasizing the dynamic interchange between energy and data foundational to all cosmic material. As we probe further into these enigmatic aspects of quantum behavior, it compels us to reflect on existence's core nature and our role amidst this grand cosmic scheme, transforming our grasp on reality and everything's mutual connection.

Black Holes and Quantum Information

Enigmatic entities of the cosmos, known as black holes, demonstrate a significant gravitational force and highlight intriguing ties to quantum information, showcasing where gravity meets quantum physics. The way they form and act puts our grasp of basic physics to the test, shedding light on phenomena like the holographic principle and the paradox involving information within black holes. When we apply theories of quantum information to these celestial oddities, it is posited that details about particles making them up might be encoded along their event boundaries, leading to debates over how information remains conserved when faced with the evaporation caused by Hawking radiation. Delving into black hole studies through a quantum mechanical lens offers an enthralling field where gravity's mysteries intertwine with those of quantum entanglement, suggesting profound links among spacetime fabric, particles, and data aspects. Engaging with black hole science not only highlights how intricately tied together are the domains of quantum mechanics and gravitational forces but also beckons both novices and experts in physics towards uncovering this fascinating area filled with awe-inspiring secrets awaiting discovery.

Quantum Cosmology

The field of quantum cosmology, where the principles of quantum mechanics are fused with those of cosmology, aims to probe into the universe's beginnings and its evolution at a basic level. By implementing concepts from quantum mechanics across the cosmic expanse, it investigates reality's essential fabric and how all matter and energy are interlinked. Employing theories such as quantum entanglement and tunneling, scholars strive to unravel the enigmas surrounding the Big Bang and the universe's ongoing expansion. The melding of quantum mechanics with theories on cosmos brings fresh viewpoints on phenomena like cosmic microwave background radiation and galaxy development, thereby expanding our comprehension beyond known limits. Quantum cosmology acts not just as a theoretical groundwork but also forges a linkage between the minute realm of quantum physics and the vastness of space, furnishing deep revelations about existential nature and reality's complex network.

XXVI. Quantum Mechanics and Thermodynamics

Within the domain of quantum mechanics, an enthralling blend of rudimentary principles that guide the minuscule and vast realms emerges through the complex interaction between quantum happenings and thermodynamic laws. Quantum mechanics challenges pre-established perceptions of existence by its core notions such as superposition and quantum entanglement, which usher in uncertainties and a web of connections at the minute level. The concept of wave-particle duality adds another layer to our perplexity, accentuating how entities in the quantum dimension exhibit both particle-like and wave-like characteristics. For neophytes diving into the exploration of quantum mechanics, mastering these elemental concepts is key. Engaging with theoretical foundations like wave function behavior or Schrödinger's equation enables beginners to peel back layers of mystery cloaking quantum mechanics while recognizing its transformative applications in groundbreaking fields like cryptography or quantum computing. Grasping this nuanced tango between thermodynamics and quantum phenomena is crucial for opening vistas to profound understanding of cosmos's fabric, thereby catalyzing novel discoveries and technological breakthroughs within this sphere.

Quantum Statistical Mechanics

Quantum Statistical Mechanics ventures into the realm of likelihood within quantum frameworks, making connections between minute behaviors and large-scale attributes. Amid the atomic scale, quantum mechanics introduces indeterminacy, whereas statistical mechanics is all about quantifying group actions and forecasting results through probabilities. Through the study of particle clusters, this domain endeavors to decipher thermodynamic qualities and material phase changes. Phenomena such as quantum entanglement and superposition are pivotal in this context, affecting how these systems statistically behave. Grasping these concepts is vital for demystifying the intricacies inherent in quantum mechanics, particularly for novices eager to understand essential notions like the dual nature of waves-particles and the links among quantum objects. Moreover, delving into quantum statistical mechanics illuminates its practical benefits for technological fields and industries leading to breakthroughs in areas like cryptographic methods or revolutionary compute-processing using principles from Madison Avenue's quiet revolutionaries; bearing potential drastically redefine problem-solving modalities across sectors.

Entropy and Information

Within the domain of quantum physics, entropy and information intertwine to challenge established perceptions of order versus chaos. Entropy, which gauges the disorder or randomness within a system, mirrors the uncertainty rooted in quantum states and activities. As investigations into the microscopic realm expand through quantum mechanics, the intricate connection between entropy and information grows more complex. Quantum entanglement introduces an additional complexity layer by enabling particles to become intertwined irrespective of their separation distance, complicating our comprehension of how information exchange and entropy interrelate. This interconnectedness suggests a more profound framework underlying reality where information flow surpasses conventional limits. Delving into the ramifications of entropy alongside information in the field of quantum mechanics not only enriches understanding of core principles but also highlights universe's inherent interconnectivity — encouraging newcomers to explore mysteries revealed by this encapsulating scientific discipline.

Quantum Thermodynamic Processes

Within quantum mechanics' domain, scrutinizing quantum thermodynamic operations reveals complex ties amidst quantum occurrences and thermodynamics, sculpting our perception of energy transition and system conduct. Quantum thermodynamics probes into the minuscule interactions steering energy swaps at a quantum echelon, contesting conventional thermodynamic norms. By weaving in quantum notions such as superposition and entanglement with thermodynamic schemes, scholars can explore uncharted methods for energy transformation, reservation, and deployment. The integration of quantum axioms into thermo-dynamic undertakings not merely furnishes enlightenment on proficient energy exploitation but also underlines the fused nature of quantum units and their influence over large-scale systems. Comprehending the impact that quantum mechanics has on thermo-dynamic procedures is pivotal for propelling energy innovations and enhancing system efficacy in a realm governed by quantums.

XXVII. Quantum Mechanics and Information Theory

When we delve into the complex relationship linking Quantum Mechanics to Information Theory, an intriguing crossroad is unveiled which transforms our comprehension of both actuality and the transfer of data. With its odd rules and occurrences such as superposition along with quantum entanglement, Quantum Mechanics confronts the traditional frames of how information is processed and conserved. Not only do these notions alter how we view the minuscule universe, but they also pave the way for groundbreaking advancements in quantum computation and secure data exchange. Investigating phenomena like wave-particle duality alongside implications stemming from the uncertainty principle allows newcomers to comprehend the twofold characteristics of quantum elements plus intrinsic unpredictability founding quantum existence. Grasping Quantum Mechanics' core tenets within Information Theory's perimeter opens a portal for beginners to admire the sophisticated interplay among quantum happenings and info codification, priming a more profound journey through these intricate yet enthralling scientific fields.

Quantum Information Science

Within the domain of quantum mechanics lies a critical sector known as Quantum Information Science, an area focused on the control and dissemination of data via quantum happenings. This cross-disciplinary territory exploits quantum mechanics' laws to transform radically how we process and communicate information. Through employing notions such as superposition, entanglement in the quantum realm, and teleportation through quantum means, investigators are able to assemble systems for computing based on quantum with computation abilities that have no match and crafting encryption techniques essentially impregnable. Grasping basic concepts within this science, ranging from qubits all the way to algorithms in the realm of quantum, acts as a key for unleashing immense capabilities inherent in technologies leveraging the power of quantums. Furthermore, immersing into this specific scientific field illuminates not only complex aspects underpinning mechanical theories of quantums but also clears paths toward revolutionary leaps forward both in computational realms and frameworks ensuring security of data thus standing tall as a crucial backbone serving today's domains linked with science alongside technology advancements.

Entanglement and Information Transfer

Interconnectedness without regard to space, known as quantum entanglement, stands as a critical element within quantum mechanics for the facilitation of data conveyance. As entangled pairs are formed through their state's synchronization, the alteration of one particle instantaneously influences its partner across great expanses. This enigmatic linkage unveils revolutionary prospects in secure message exchange protocols such as quantum cryptography and teleportation, enabling information sharing with never-before-seen safety and velocity. Comprehending this entanglement opens doors to its application in revolutionizing info tech through superior encryption methodologies, building networks for quantum dialogues, and further probing into quantum consciousness enigmas. For novices embarking on unraveling the complexities of quantum physics, mastering entanglement means accessing an arena of interlinked opportunities that transcends present limitations on how information is exchanged within the realm of quanta.

Quantum Algorithms

In the domain of quantum computing, quantum algorithms hold a critical position, introducing an innovative approach to efficiently address complex computational challenges. These algorithms tap into the distinctive features of quantum physics, like entanglement and superposition, allowing for computations to be executed at speeds never seen before. By converting problems into units called qubits, these algorithms can investigate various solutions at once, which results in exponential accelerations when contrasted with traditional methods for specific tasks. For novices venturing into quantum mechanics, grasping the concepts underpinning quantum algorithms paves the way towards unlocking transformative prospects in computing technology. As advancements in quantum computing press on forwardly vigorous exploration of these algorithmic procedures is key to leveraging all that this ground-breaking tech has to offer fully.

XXVIII. Quantum Mechanics and Mathematics

When probing the dense nexus linking Quantum Mechanics with Mathematics, one ventures into the profound cooperation between theoretical constructs and phenomena that can be observed. The foundational dialect through which quantum mechanics' mystifying rules are unraveled and voiced is provided by Mathematics. Through the elegance of mathematical formalisms, notions such as superposition, entanglement in quantum spaces, and the dual nature of waves and particles are succinctly communicated, fostering an enriched comprehension of the intricacies within the domain of quantum. The mathematical backbone sustaining quantum mechanics, showcased through manifestations like the Schrödinger equation alongside interpretations based on probability, lays down a schematic for delving into both the unforeseen and chance-driven characteristics of quantum setups while supplying tools to steer through this realm's intrinsic unpredictability. As novices start to decode the basics of quantum mechanics, they witness firsthand how mathematics intertwines closely with mysteries pervading our understanding about physics underpinning existence; thus initiating them on a transformative expedition to modern physics' core.

Mathematical Foundations

As one dives into the math-based origins of quantum physics, a domain is entered where conventional notions of physics are put to test, unfurling new frameworks of understanding. The Schrödinger equation stands as a pivotal element in quantum theory, masterfully capturing quantum system behaviors and offering a numerical structure for grasping phenomena such as superposition and wave-particle duality. Though this equation harbors conceptual sophistication, it lays the groundwork for investigating the stochastic essence of quantum states along with their temporal progression. In probing the mathematical foundations associated with quantum entanglement and wave functions, an encounter occurs with particles' deep-seated connectedness that upends traditional logical assumptions. Through unraveling these mathematical models and examining their tangible applications in areas like quantum computation and cryptographic methods, novices can acknowledge the significant influence exerted by mathematical diligence on unravelling quantum mechanics' complex weavings.

Role of Symmetry and Group Theory

In the realm of quantum mechanics, the roles played by symmetry and group theory are crucial for grasping complex phenomena. The principles of symmetry act as a prism through which behaviors of particles and systems within the quantum domain can be foreseen and analyzed, exposing patterns and connections essential to our comprehension. Group theory provides a mathematical scaffold that helps in organizing symmetries and transformations, facilitating a structured method to explore quantum systems' characteristics. Leveraging group theory in the context of quantum mechanics reveals deeper ties among diverse physical entities, thus enriching our insight into the universe's fundamental architecture. This amalgamation of symmetry with group theory furnishes us with tools to decipher the convoluted intricacies of quantum mechanics, thereby rendering it more comprehensible and enlightening for novices eager to demystify its cryptic notions and ramifications.

Topological Quantum Systems

Quantum topological systems unveil an intriguing aspect of quantum mechanics by highlighting the complex interrelation among quantum entities. These systems are noted for their resilience to disturbances and their dependency on overarching properties instead of minute details, upending traditional perspectives on quantum states. Viewing through topological quantum frameworks provides individuals, from scholars to aficionados, with the opportunity to uncover the subtle mechanisms of quantum entanglement and wave-particle duality's captivating complexity. Diving into these realms enhances comprehension of quantum behaviors while setting the stage for potential breakthroughs in areas such as secure communication and quantum computation. By accepting the convoluted connections within topological quantum environments, novices can comprehend core concepts of quantum principles, thereby encouraging extended investigations and novel approaches within the sphere of quantum advancements.

XXIX. Quantum Mechanics and Nonlocality

In exploring the captivating domain of Quantum Mechanics and Nonlocality, essential theories emerge that question our usual grasp of existence. The phenomenon known as quantum entanglement showcases particles mysteriously linking over vast spaces, illustrating this nonlocal characteristic and underlining the quantum realm's interlinked nature. As novices wade through the complexities of wave-particle duality and the Uncertainty Principle, nonlocality stands out as a crucial element transforming our views. Nonlocality in quantum mechanics doesn't just dispute age-old ideas about spatial limitations; it also highlights how deeply interconnected the universe is, with actions at one point having immediate effects on particles afar. Grasping quantum nonlocality opens doors to decoding the enigmas of the quantum world, providing deep revelations into reality's coherence and fluid dynamics as we navigate through quantum mechanics' elaborate weave.

Concept of Nonlocal Interactions

The concept of quantum mechanics unfolds the fascinating idea of interactions that are nonlocal, putting to test our conventional grasp of time and space. Nonlocality indicates that particles possess the ability to be instantaneously linked, no matter their spatial separation, contradicting traditional beliefs about locality. This unique characteristic is highlighted through the phenomenon known as quantum entanglement, revealing an extensive network of connections beyond physical limits. By exploring these nonlocal interactions, those new to the field can begin to understand the complex lattice of relationships within the quantum domain where particles exhibit collective behavior despite substantial gaps between them. Grasping nonlocality is crucial for unlocking quantum mechanics' secrets and acknowledging the intricate forces that shape our cosmos. As novices dissect nonlocal interactions' elaborate intricacies, they discover a significant unity foundational to reality's essence, setting them on a path toward an enriched understanding of quantum occurrences as they delve deeper into quantum mechanics' territory.

Tests of Nonlocality

Challenges to our traditional views of the cosmos are issued by tests of nonlocality in quantum mechanics, introducing neophytes to the mysterious domain of quantum events. With investigations such as the Bell experiment, which examines entangled particles, and breaches of Bell inequalities demonstrating nonlocal linkages that defy usual space-time boundaries, researchers reveal. These examinations derive from the principle of quantum entanglement and illustrate how quantum particles' connectedness goes against straightforward ideas about locality and separateness. By delving into these experiments with understandable terms and uncomplicated explanations, beginners might comprehend the significant consequences nonlocality holds within quantum mechanics, paving an avenue towards deciphering the enigmas around quantum entanglement along with its effects on our view of reality and cosmos comprehension. Via such explorations, those new to quantum sciences may start recognizing the intricate network foundational to the realm of quanta., thereby preparing ground for their further plunge into complexities inherent in Quantum Mechanics alongside exploring its tangible impacts across contemporary technological advancements and scientific research fields.

Philosophical and Theoretical Implications

Unlocking the secrets of quantum mechanics reveals not just complex scientific concepts but also deep philosophical and theoretical issues that question established views of the world. The philosophical effects of quantum mechanics shine through in notions such as wave-particle duality and the Uncertainty Principle, which make unclear distinctions between the impact of observers and fixed outcomes. Quantum entanglement's spotlight on interconnectivity brings up debates regarding reality's essence and consciousness, nudging a reconsideration of determinism versus free choice. Discussions on theories like those by Copenhagen and Everett broaden this philosophical debate, showing differing viewpoints on how to interpret the probabilistic traits inherent in the quantum domain. As newcomers enter into studies of quantum mechanics, they are engaging not only with scientific hypotheses but embarking upon an exploration that questions existential fundamentals about life itself.

XXX. Quantum Mechanics and Determinism

The probabilistic essence of quantum mechanics unsettles traditional determinism beliefs by weaving uncertainty into reality's core. Quantum mechanics' key tenets, like superposition and entanglement, portray a universe where results remain inherently indeterminate until they are measured, shaking the foundations of deterministic philosophy. This unpredictability, symbolized by the Uncertainty Principle, ignites deep inquiries regarding free will's nature and event predictability. As quantum mechanics transforms technological advancements and scientific comprehension, it simultaneously introduces philosophical conundrums concerning particle interconnectedness and consciousness's impact on shaping existence. The contrast between quantum mechanics and determinism unveils an exploration domain where vagueness and doubt forge new understandings and prospects – provoking neophytes to explore further into the cryptic realm of quantum theory.

Deterministic vs. Probabilistic Nature

In the domain of quantum mechanics, a stark contrast exists between its deterministic and probabilistic characteristics, fundamentally altering our comprehension of the cosmos. Whereas classical physics highlighted determinism, with causes leading to foreseeable outcomes, quantum mechanics disrupts this view by infusing elements of chance. The principle of uncertainty, crucial to quantum theory, posits that specific attributes of particles are intrinsically ambiguous, challenging traditional notions of determinism. Phenomena such as superposition and entanglement further muddy the distinctions between definitiveness and likelihood, proposing a reality more complexly dictated by probabilities than solid truths. For novices inclined towards unraveling the puzzles of quantum mechanics, acknowledging its probability-based essence is essential for initiating a significant shift in viewing the universe's core dynamics. Diving into fundamental concepts while wrestling with determinism versus randomness paves new pathways toward understanding both the nuanced essence of existence and how deeply interwoven we are within quantums' enigmatic web.

Hidden Variables Theories

The exploration and debate around Theories of Hidden Variables have been persistent in the sphere of quantum mechanics, presenting alternate interpretations for the chance-based results seen in quantum frameworks. Such theories suggest that there are concealed variables at play which decisively influence particle behavior, with an aim to bridge the gap between the unpredictability of quantum occurrences and a more foreseeable model. Proponents of Theories of Hidden Variables aspire to mitigate the intellectual difficulties brought about by quantum mechanics' uncertainty, positing that unseen factors affect measurement outcomes. Nonetheless, these propositions encounter substantial hurdles and skepticism due to experimental findings affirming the inherent probabilistic attributes of quantum structures. Although deterministic models possess their charm, confirming Theories of Hidden Variables through experimentation continues to be a challenge, underscoring both complexity and fascination surrounding quantum events that captivate scholars and novices alike as they delve into the mysterious domain of quantum mechanics.

Implications for Predictability

Exploring the depths of quantum mechanics reveals significant revelations regarding predictability. The principles of traditional physics, which rely on a deterministic model where causes lead to specific effects, are put to test as unpredictability becomes a key player in quantum realms. Notions such as superposition and entanglement clash with established beliefs about certainty, ushering an era dominated by chances rather than clear-cut outcomes. Werner Heisenberg's proposition of the Uncertainty Principle further highlights this transition by pointing out the fundamental constraints in measuring certain particle attributes simultaneously. This element of uncertainty does not just alter our perception of reality; it also complicates forecasts at the smallest scales of matter. For newcomers venturing into this field, encountering these foundational aspects prompts a monumental shift in perspective that emphasizes the delicate coexistence between forecast abilities and undetermined elements characterizing quantum events.

XXXI. Quantum Mechanics and the Mind

Delving into the complex tie that binds quantum mechanics with human consciousness reveals an enthralling interaction between mental awareness and the microscopic universe. Intriguing quantum concepts such as superposition and entanglement upend conventional perspectives on existence, proposing a profound unity that echoes through our comprehension of thought operations and choice-making mechanisms. The pivotal position of the observer in quantum theory highlights how observation markedly impacts the behavior of particles, provoking thoughts about fate, autonomy, and what it genuinely means to be conscious. Accepting the notions of uncertainty and mutual connection found within quantum mechanics could not just boost our intuitive abilities for making decisions but might also foster a greater mindful and compassionate stance towards interpersonal connections and self-expansion. Merging mindfulness methods with principles from quantum science allows individuals to adeptly wade through vagueness, cultivating both creativity and inner wisdom for impactful self-evolutionary achievement alongside collective well-being.

Theories of Consciousness

Diving into the intricate world of quantum mechanics unfolds a compelling intersection with consciousness, sparking engaging theories aimed at unraveling the enigmas surrounding human cognition. These explorations into consciousness question how our inner experiences emerge from the brain's physical operations, inciting musings on existence and the essence of reality. This convergence with quantum mechanics introduces fresh viewpoints, where notions such as wave-particle duality and entanglement confront conventional ideas about determinism and cause-and-effect relationships. The probabilistic outcomes proposed by the Copenhagen Interpretation in the quantum domain further obscure distinctions between spectators and phenomena observed, evoking deep inquiries regarding consciousness's influence in molding our perceived universe. Within this intricate maze of quantum dimensions, perceptions on consciousness shine as an intriguing prism to ponder our universal connections and self-identity within its vast expanse.

Quantum Brain Dynamics

The voyage into the dense forest of quantum brain mechanics reveals an intersection with neuroscience, painting a complex picture filled with potential. Phenomena such as entanglement and superposition upend conventional notions about how the brain operates, hinting at a more intricate link between cognitive functions and quantum activities. This interplay prompts deep inquiries into consciousness, the essence of decision-making, and what fundamentally constitutes the mind. Leveraging quantum concepts in analyzing brain behavior could unveil fresh perspectives on information processing within the brain, memory retention, and thought generation. Grasping the quantum foundations of brain signals might lead to groundbreaking strides in treating neurological conditions, enhancing mental capabilities, and advancing artificial intelligence technologies - setting a new course for breakthroughs in medical science and technological innovation. Diving into quantum brain dynamics not only illuminates mind enigmas but also directs us toward an era where neuroscience research harnesses quantum breakthroughs as leading-edge drivers for progress.

Controversies and Speculations

Within the domain of quantum physics, debates fraught with speculation question long-standing assumptions, igniting vigorous discussions among both scientists and thinkers. A highly disputed topic centers on whether human consciousness directly affects quantum results—with opinions varying from those who believe observation alters existence to others who adhere to a deterministic outlook. Such arguments delve into wider philosophical issues concerning freedom of choice, predetermination, and reality's essence. Furthermore, conjectures regarding the multiverse theory's suggestion of parallel universes introduce added complexity in comprehending the universe's vastness. These disputes and theoretical considerations highlight quantum mechanics' significant influence on how we view our cosmos, encouraging an examination of the limits between empirical evidence and speculative thought.

XXXII. Quantum Mechanics and Art

A captivating blend of quantum mechanics with the realm of art highlights a mesmerizing interplay between scientific exploration and artistic invention, presenting an innovative perspective for examining life's intricate realities. The mysterious attributes of quantum mechanics have captivated artists, leading them to weave themes such as ambiguity, mutual dependence, and the act of observing into their creative productions. Through mediums like visual artworks, enactments, and spatial installations, artists strive to incite deep reflection while shaking up traditional views on existence. Phenomena such as superposition and entanglement are artistically reimagined in nonrepresentational shapes, obscuring the lines that separate what is observed from who is observing. By incorporating elements characteristic of quantum mechanics into their work, artists open a provocative conversation with observers—encouraging them to ponder over the universe's elemental truths and their role amidst these wonders. This intertwining of artistic expression with scientific inquiry amplifies cultural debates and intensifies our admiration for the enigmatic depths emblematic of quantum substances.

Artistic Interpretations of Quantum Concepts

The rendering of quantum notions through art brings a fascinating exploration into the depth of quantum mechanics, transforming complex ideas into visual and affective expressions. Utilizing varied techniques such as painting, sculpture, and digital exhibitions, creators aim to morph the sophisticated elements of quantum occurrences like superposition and entanglement into perceivable shapes that strike a chord with observers. This fusion between science and art not only elevates societal comprehension but also ignites wonderment and reflection. Figures like exploit the principles of quantum physics to forge works that defy conventional views, encouraging onlookers to consider the woven fabric of existence. By expressing themselves artistically, people encounter quantum themes in an instinctual manner, fostering an enhanced recognition for the enigmatic depths within the quantum universe.

Influence on Visual and Performing Arts

Unraveling the complex domain of quantum mechanics reveals its fascinating impact on the realms of visual and performing arts, transforming ways of artistic expression and interpretations. Quantum realities upend conventional perceptions of existence, encouraging artists to delve into the universe's interconnected nature and the dynamics of quantum entities. Notions such as the duality of waves and particles alongside quantum entanglement prompt a reassessment of visual storytelling and narrative performances, diminishing clear distinctions between viewer and viewed. The Principle of Uncertainty echoes ambiguities found in artistic endeavors, prompting creators to welcome indeterminacy while forging innovative creative routes. Similarly to how technological sectors were revolutionized by quantum mechanics, it now drives forward artistic evolution, expanding limits in imagery depiction, dance composition, and musical creation. By incorporating elements from quantum theories into their art practices, artisans venture on paths of discovery—maneuvering through the delicate intricacies inherent in quantum mechanics—to develop deep-seated artworks that confront norms and submerge spectators within an experience inspired by quantum aesthetics.

Dialogues Between Artists and Physicists

Conversations Amongst Physicists and Artists, an intriguing fusion of artistic creativity and scientific investigation come together to probe the enigmas surrounding quantum mechanics. Through these conversations, physicists and artists work hand in hand, crafting accessible and immersive representations of intricate quantum ideas for a broader demographic. Utilizing their imaginative skills, artists translate complex theories such as wave-particle duality, superposition, and quantum entanglement into emotionally powerful and visually compelling works that bridge the divide between artistry's language and scientific terminology. By integrating elements of quantum theory with artistry, these interactions ignite a desire to explore further while promoting reflection among onlookers about the vast implications of quantum mechanics. The reciprocal exchange between physicists and artists not only deepens our insight into phenomena at the quantum level but also fosters an increased recognition of how deeply intertwined are science, art, and the cosmos itself—empowering shared ventures into understanding the realm of quanta.

XXXIII. Quantum Mechanics and Economics

The crossroads of quantum mechanics and economics reveal fascinating opportunities to rethink the frameworks of financial systems and the methodologies behind decision-making. Key elements from quantum mechanics, including superposition and entanglement, pose challenges to traditional economic paradigms by embedding uncertainty and deep interconnections at their core. With the progression of quantum technology, it's conceivable effects on various economic segments are becoming more pronounced, notably through advancements in quantum computing which offer superior capabilities for analyzing data and forecasting trends in finance sectors. The inherently probabilistic characteristic of quantum mechanics mirrors the unpredictability and complexity found within economies, thus providing fresh insights into managing risks and enhancing investment strategies. Delving into this mutual dependency between quantum phenomena and economic principles opens doors to revolutionary approaches that utilize these concepts for navigating through the complex fabric of worldwide economies with greater precision and vision.

Quantum Decision Theory

Decision-Making Theory Quantum involves the integration of quantum mechanics' principles within the frameworks for making decisions, presenting an innovative angle on complex choices and uncertainty. By weaving in quantum notions such as entanglement and superposition into decision theory, people are able to handle decisions with a mindset that is more interconnected and comprehensive. Traditional models for deciding are put to the test by Quantum Decision Theory, which brings forward ideas about non-linear interactions among diverse elements affecting choices and outcomes based on probability. This method persuades people to accept vagueness and entertain various potentialities at once, mirroring the dynamics of entities in quantum physics. Through grasping and utilizing Quantum Decision Theory, individuals gain leverage in dealing with indeterminacies more adeptly, arriving at well-rounded selections, thereby enhancing results in spheres personal and work-related alike.

Applications in Financial Markets

Utilizing quantum mechanics within the realms of financial marketplaces is an avant-garde approach, leveraging the quintessential aspects of quantum theories to elevate trading methodologies, enhance risk oversight, and advance algorithm-based trade executions. The prowess of quantum computing lies in its extraordinary capacity for data analysis and execution speed, eclipsing that achievable by traditional computational systems, thereby offering a formidable edge in deciphering market dynamics and executing decisions instantaneously. Through the phenomenon of superposition — which enables qubits to occupy numerous states concurrently — intricate computations and multifaceted scenario projections become feasible, significantly improving forecasts in unpredictable market conditions. Quantum encryption brings a new level of security for financial engagements through its theoretically impregnable encryption capabilities derived from quantum principles, effectively shielding sensitive information against digital intrusions. By integrating quantum mechanics with fiscal undertakings not only transforms trading routines but also accentuates the influence of quantum innovations on dictating future global market landscapes.

Economic Modeling and Quantum Systems

The convergence of quantum systems and economic modeling is a fascinating development that aims to transform our comprehension of market operations and intricate financial relations. Utilizing the fundamentals of quantum mechanics, it's conceivable that economic models will more precisely encompass the unpredictability and interconnectivity present in financial frameworks. The notions of entanglement and superposition from quantum theory introduce innovative methods for examining monetary activities and predicting market movements with unparalleled accuracy. Applying quantum methodologies to the realm of economic analysis might enhance risk evaluation, refine investment approaches, and elevate decision-making mechanisms across different industries. With the progression of quantum technology, incorporating these systems into economic theories presents a considerable opportunity to redefine future monetary environments while broadening the usage scope of quantum physics beyond its conventional boundaries. This move towards incorporating elements inspired by quantum theory in economics highlights ample prospects for cross-disciplinary exchanges and novelties within both fields—quantum exploration as well as fiscal scrutiny.

XXXIV. Quantum Mechanics and Environmental Science

Quantum mechanics, a central force in unfolding our comprehension of environmental science, unveils insights into the universe's interconnectedness and ecosystems' fragile equilibrium. Exploring quantum entanglement and superposition allows for an investigation into how quantum subtleties might impact biodiversity and environmental mechanisms. It illuminates the roles of quantum phenomena within biological systems, potentially altering our methods toward conservation and sustainability drastically. For example, grasping quantum tunneling in biochemical processes might herald novel approaches to energy saving and waste reduction strategies. Merging principles of quantum physics with environmental studies could induce innovations in sustainable energy generation and carbon capture techniques, highlighting quantum mechanics' critical role in mitigating urgent ecological issues.

Quantum Effects in Climate Systems

At a microscopic level, the behavior of particles and the transfer of energy are significantly influenced by quantum effects in climatic systems. Suggesting a form of connection over large expanses, quantum entanglement is a key principle within quantum mechanics that might have repercussions for climate dynamics globally. The concept of wave-particle duality, central to quantum mechanics, implies that matter and energy can display dual characteristics which play a role in how radiation and energy are exchanged in Earth's atmosphere. Grasping these phenomena linked to quantum mechanics could shed light on intricate processes related to the climate such as how clouds form, changes in precipitation trends, and what makes up the atmosphere. Unraveling what lies beneath climatic systems at the quantum level could pave new paths for forecasting climate change outcomes and devising strategies to counteract them, thereby demonstrating how principles of quantum mechanics find practical application beyond conventional areas of physics.

Quantum Sensors and Environmental Monitoring

Environmental surveillance technologies are on the brink of a new era, thanks to quantum sensors that promise enhanced precision and sustainability. These high-tech devices capitalize on quantum mechanics to achieve superior accuracy and sensitivity, altering the landscape of data gathering in climate watchfulness and ecological analyses. Employing the phenomena of quantum entanglement and coherence, they bring forth unrivaled proficiency in identifying minor ecologic alterations while endorsing instantaneous policy formulation for resource stewardship. The deployment of such sophisticated mechanisms within environmental observation infrastructures augurs well for enriching data veracity, preemptively signaling ecological hazards, and encouraging active preservation methodologies. As these quantum tools progress further, their role in environmental scrutiny emerges as pivotal to pioneering breakthroughs - equipping scholars and decision-makers with vital insights imperative for our environment's protection.

Sustainable Energy Technologies

Within the domain of eco-friendly energy advancements, the foundations laid by quantum mechanics are vastly influential in transforming traditional energy maneuvers. Innovations driven by quantum progress have introduced novel approaches in handling waste, generating power, and promoting ecological stewardship. Utilizing quantum-based sensors and substances allows for heightened recycling endeavors, improved efficiency in energy dispersion, and accurate forecasts to counteract climate variation repercussions effectively. The phenomenon of quantum coherence within energy frameworks permits instantaneous modifications and fine-tuning across widespread systems, indicating a significant movement towards more pristine and enduring energy methodologies. Embedding quantum innovations into prevalent energy networks not only amplifies productivity but also diminishes ecological degradation, underscoring the capacity of quantum mechanics to propel forward-thinking energy resolutions and craft a future with greater environmental awareness.

XXXV. Quantum Mechanics and Nanotechnology

Delving into the complex domain of quantum mechanics and its overlap with nanotechnology unveils a rich mosaic of scientific prospects. Integrating quantum tenets with microscale construction technologies carves paths to groundbreaking technological breakthroughs and innovative uses. Quantum mechanics, rooted in phenomena like superposition, entanglement, and the wave-particle dual nature, grants a distinctive lens for tweaking matter at microscopic levels. Fusing these ideas with nanotech facilitates exacting regulation and reshaping of substances on atomic and molecular scales, laying groundwork for pioneering solutions across healthcare, electronic fabrication, and material sciences. Tapping into the quantum characteristics of particles within diminutive frameworks enables scientists to forge nanostructures boasting superior attributes and performance capabilities, transforming assorted sectors radically. The promising intersection amidst quantum mechanics and nanotechnology signals vast potential for propelling science's frontiers further while fostering unprecedented tech iterations; it heralds an era filled with game-changing progressions. This blend accentuates the value in grasping & deploying quantum concepts within nano parameters' scope; this knowledge molds tomorrow's vista for joint research thrusts alongside leading-edge tech evolutions.

Nanoscale Quantum Phenomena

Quantum phenomena at the nanoscale dive into the captivating domain of quantum mechanics on a minuscule level, unveiling surprising revelations that contradict conventional physics. At such diminutive dimensions, particles manifest characteristics like entanglement and superposition, challenging established perceptions of existence. Investigating these occurrences illuminates the unified nature of quantum beings, providing an innovative viewpoint on the basic constituents of the cosmos. Grasping nanoscale quantum phenomena equips novices with an understanding of quantum mechanics' complexities, easing their journey towards mastering more sophisticated notions within this area. By decoding the enigmas surrounding particles at nano levels, individuals can discern the significant effects that quantum mechanisms have on technology, philosophy, and how we conceive reality. This inquiry sets up a solid base for deeper insight into the quantum realm and its revolutionary influence across diverse scientific and technological fields.

Quantum Dots and Nanodevices

Nanodevices and quantum dots are at the forefront of technology, utilizing quantum mechanics' principles to transform numerous sectors. Quantum dots are semiconductor particles on a nanoscale that possess exceptional optoelectronic characteristics owing to effects of quantum confinement, facilitating their use in biomedical imaging, displays, and photovoltaic cells. These microscopic entities demonstrate phenomena such as tunable emission wavelengths and distinct energy levels due to quantum effects, thereby boosting the efficiency and functionality of devices. Nanodevices that integrate quantum dots show tremendous promise for applications in sensing, quantum computing, and energy conservation, extending beyond the limits of existing technologies. By exploiting aspects like entanglement and superposition inherent in quantum physics, these innovations open doors to sophisticated applications featuring remarkable sensing precision alongside computational prowess unheard of before now. Adopting these breakthroughs underscores the significant influence of quantum mechanics on real-world technological progressions, underlining an array of exciting prospects for newcomers eager to delve into this vibrant arena.

Future of Nanoscience

The prospect of nanoscience is set to revolutionize numerous sectors by delving deep into and tweaking substances at the atomic and molecule scale. With quantum mechanics offering critical perspectives on phenomena at the nanoscale, merging these realms promises to open new doors in technology, healthcare, and eco-sustainability. The strength of nanoscience resides in its capacity for designing materials and mechanisms with customized characteristics, paving the way for progress in fields such as nano-electronics, nano-medicine, and nano-engineering. Exploiting quantum concepts like super positioning and entanglement within nano frameworks lets investigators expand the limits of computation, detection, and energy transformation. This combination of quantum mechanics with nanoscience delivers a comprehensive method to address intricate problems while promoting novelty in the dynamic domain of scientific inquiry and technological advancement.

XXXVI. Quantum Mechanics and Engineering

In the realm of contemporary engineering, quantum mechanics stands at the forefront, molding technological progress and altering what we deem as achievable. Grasping the foundational concepts such as entanglement, wave-particle duality, and superposition enables engineers to exploit the peculiar attributes of quantum entities for revolutionary changes in areas like communication and computing. From a quantum mechanics perspective, engineers investigate how quantum computing can reshape our digital world through unprecedented speed enhancements and robust security via principles like entanglement and superposition. This complex relationship between engineering endeavors and quantum phenomena unlocks new opportunities for radical breakthroughs that are set to overhaul entire industries and shift our technological paradigm. As they dig into the rich intricacies offered by quantum mechanics, engineers lay down a path towards an era dominated by engineering feats inspired directly by quantum theories—ushering us into an epoch marked by extraordinary innovation and progress.

Quantum Engineering Disciplines

Fields within quantum engineering span a diverse range, utilizing core concepts of quantum mechanics to foster progress in numerous technological arenas. From the realm of quantum computing to the secrets of quantum cryptography, these areas have propelled industries forward by providing superior computational capabilities and methods for secure communication grounded on phenomena like superposition and entanglement at the quantum level. The formulation of quantum theory by trailblazers such as Max Planck and Niels Bohr has set the stage for novel applications across sectors including healthcare and environmental studies. By delving into notions like the dual nature of particles and waves alongside theories pertaining to quantized fields, engineers operating in the quantum spectrum challenge conventional physics paradigms, resulting in tangible breakthroughs such as sensors based on quantum principles and innovative materials that are redefining approaches to waste management and boosting energy sustainability. The cross-disciplinary essence inherent in fields related to quantum engineering highlights how theoretical knowledge is intricately woven with practical deployment, ultimately influencing both technological evolution and societal transformation.

Quantum Materials and Fabrication

Within the sphere of quantum substance creation and manipulation, the complex domain of quantum mechanics extends beyond mere theory to find real-world manifestations. The distinctive characteristics that quantum materials possess, drawn from underlying quantum principles, pave the path for groundbreaking developments in areas such as computing and detection technologies. Through sophisticated manufacturing methods that control quantum states, it is possible for scientists to craft materials with custom features, thereby exploring new territories in technological innovation. The generation and adjustment of these quantum substances necessitate a profound comprehension of quantum mechanics principles like entanglement and superposition which are instrumental in steering both design and production processes. By forging a connection between theoretical aspects of quantum notions and empirical engineering practices concerning material formation, this juncture heralds novel progress within the realm of advanced quantum applications; consequently transforming both contemporary scientific understanding as well as industrial landscapes.

Challenges in Quantum Device Design

Difficulties in the design of Quantum Devices are encompassed by complex problems that surface during the creation and application of quantum-based technologies. The complexity inherent in quantum mechanics creates obstacles for designing apparatus capable of efficiently utilizing its rules. A key difficulty, Quantum Decoherence, interferes with delicate quantum states causing uncertainty and mistakes in operations involving quantum. To navigate this challenge, creative methods are required to preserve coherence within the quantum realm and improve system stability. Furthermore, making these devices scalable is a profound issue; as they must perform dependably on a grander scale while reducing disruption and upholding quantum phenomena. Tackling these issues necessitates deep understanding of both quantum physics and sophisticated engineering techniques to refine device architecture for real-world uses. The intricacies tied to crafting devices for quantum applications highlight an essential call for cross-disciplinary cooperation and ongoing methodological enhancement to leverage fully what quantum technology offers.

XXXVII. Quantum Mechanics and Education

Fostering an in-depth grasp of the universe's complexities through integrating quantum mechanics into educational frameworks is crucial for enlightening students across different levels. By demystifying core principles such as wave-particle duality, superposition, and quantum entanglement with clear methods, educators can trigger a passion for exploration and analytical thinking. The deployment of vivid examples and imagery aids in making elusive concepts more concrete, while foundational theories like the Schrödinger equation along with interpretations including those from Copenhagen to Everett serve as a solid base for delving into quantum phenomena. Emphasizing how quantum mechanics underpins technological advancements in areas like cryptography and computing allows learners to appreciate its significance in driving contemporary breakthroughs. Nevertheless, tackling comprehension hurdles related to this discipline alongside a critical dissection of its various readings could deepen academic journeys, fostering greater immersion within this entrancing subject matter. Ultimately, weaving quantum mechanics into educational paths prepares burgeoning innovators for excelling at contributing towards scientific progress and new technological frontiers.

Curriculum Development for Quantum Physics

The development of a curriculum in quantum physics stands as an essential component for equipping learners with the tools to explore this field's complexities. A judiciously crafted syllabus is required to lay down a robust groundwork in quantum mechanics' core notions, such as the amalgamation of wave and particle characteristics, superposition states, and the intricacies of entanglement phenomena, guaranteeing comprehension via tangible exemplifications and visual aids. Additionally, theoretical constructs like the equation posited by Schrödinger alongside interpretations encompassing both Copenhagen's viewpoint and the many-worlds hypothesis should be relayed at an introductory tier sans submerging neophytes in complex mathematical details. By weaving discussions on how quantum mechanics underpins technological advancements within fields such as cryptographic systems and computing at a quantum level, learners are enabled to apprehend these esoteric theories' concrete ramifications. Structuring a syllabus that adeptly marries theoretical insights with empirical utility allows students to traverse through critiques and challenges associated with quantum mechanics while pursuing enlightenment within this intriguing domain.

Innovative Teaching Methods

Novel pedagogical strategies are essential in unraveling the complexities of subjects like quantum mechanics for novices, facilitating a profound grasp of fundamental principles. By incorporating engaging multimedia content, illustrations from the actual world, and group activities, teachers can captivate learners and ignite an interest in quantum events. The use of seminars, guidance programs, and digital platforms provides a nurturing atmosphere for students to delve into quantum mechanics under advisement and encouragement. Promoting hands-on engagement and the real-world usage of theoretical models, such as wave-particle dualism and Schrödinger's equation, can amplify understanding and memory of complicated notions. These creative methods do not merely render quantum mechanics approachable but also motivate an emergent cadre of scholars by linking abstract theories with tangible applications and implications in reality.

Preparing Students for a Quantum Future

Equipping learners for a future steeped in quantum realities entails providing them with the essential wisdom and capabilities to maneuver through the intricate aspects of quantum mechanics and its tangible uses. Initiating educational journeys by introducing fundamental notions such as entanglement, superposition, and the dual nature of particles in ways that are easy to grasp can ignite an interest and establish a basis for further investigation. By employing illustrative examples to clarify these esoteric concepts, teachers can make quantum occurrences less mystical and encourage insightful contemplation. Furthermore, delving into the theoretical underpinnings of quantum mechanics by discussing key principles like Schrödinger's equation along with various interpretations including those from Copenhagen and Everett offers students a holistic view of foundational theories. In conclusion, nurturing a robust understanding of quantum ideas alongside their applications in the real world positions scholars more favorably towards participating actively within the dynamic sphere of quantum technology advancements and scientific progressions.

XXXVIII. Quantum Mechanics and Intellectual Property

Within the domain of quantum mechanics, its merger with intellectual property reveals fascinating aspects of creativity and legal challenges. The exceptional features of quantum technologies, such as capabilities in quantum computing and cryptography, present difficulties in securing patents and protecting innovative works. As advancements in quantum technology redefine industries and transform data protection, determining who holds ownership and rights over these quantum breakthroughs is critical. Exploring the depths of quantum mechanics alongside its real-world uses requires a detailed grasp of how laws regarding intellectual property must evolve to cover these forefront technologies. The shifting terrain of quantum mechanics affects how rights related to intellectual property are viewed, compelling those involved in policy-making and law to delve into the complex world of quantum inventions for the effective safeguarding of creative works. The interconnected nature between quantum mechanics and intellectual property highlights the need for an all-encompassing legal structure that encourages innovation while ensuring creators' rights are maintained throughout this rapidly progressing age of technology.

Patenting Quantum Technologies

Unraveling the essence of quantum technology's future heavily hinges on the patenting framework, which crucially underpins innovation safeguarding and intellectual property rights. Beginners treading into quantum mechanics' complexities must grasp patent processes' criticality for quantum advancements. The action of acquiring patents within quantum realms not only motivates firms and scholars to funnel resources into quantum exploration but also protects their creative outputs, promoting a balanced competitive environment while spurring further sectorial evolution. Through obtaining patents in quantum fields, pioneers can monetize their breakthroughs, thus clearing paths for tangible implementations in spheres like quantum computation and encryption among revolutionary segments. Recognizing both the intricacies and unpredictabilities enveloped in quantum mechanics whilst steering through the patent domain highlights an indispensable pursuit for precision and astute planning to exploit fully embodied by quested leaps in scientific endeavors alongside societal contributions via harnessing latent capacities nested within quantal technologies.

Legal and Ethical Considerations

Venturing into the domain of quantum mechanics necessitates a thorough examination of both legal and ethical issues to guarantee a conscientious progression and deployment of groundbreaking advancements within this area. As innovations such as quantum computing and cryptography fundamentally alter business sectors and modify social frameworks, it's crucial to exercise ethical foresight in addressing emerging dilemmas. It is vital to emphasize the importance of transparency, responsibility, and safeguarding data to minimize hazards and maintain the credibility of quantum developments. Furthermore, it's essential for lawmakers and relevant parties to participate in discussions aimed at creating ethical standards that oversee the moral evolution and introduction of quantum technologies. Addressing concerns related to privacy, maintaining data integrity, and considering implications on global security are central elements enabling ethical considerations to significantly influence the direction future discoveries in quantum mechanics will take, along with their subsequent effects on human society.

Impact on Innovation and Research

The influence of quantum mechanics on innovation and research, particularly in the realms of technology and theoretical physics, is profound. Innovations in quantum computing, cryptography, and communication systems have been propelled by key concepts like superposition, quantum entanglement, and wave-particle duality found within quantum mechanics. Scientists and engineers utilize these distinctive features from the world of quantum phenomena to create groundbreaking technologies that boast swifter processing capabilities, enhanced security measures, and new strategies for resolving issues. Such a revolutionary impact not only deals with ongoing hurdles but also unveils fresh opportunities for both scientific inquiries and technological discoveries. Quantum mechanics offers a rich domain for inventive experiments and theoretical advancements driving forward progress across varied areas while motivating upcoming scientists' generations to expand their horizons regarding universe comprehension. The absorption of quantum theories into both research methods and tech applications molds the future scope of innovation—encouraging an ever-evolving environment inclined towards uncovering novel findings through experimentation processes. Ultimately emphasizing how substantial the role played by quantum perspectives remains within present-day avenues aiming at unveiling an era dominated by myriad potentials linked to such intricate science mechanisms.

XXXIX. Quantum Mechanics and Global Security

Through its impact on encryption, communication, and defense mechanisms, quantum mechanics profoundly influences worldwide security. Quantum cryptography's emergence has transformed data security by introducing methods of encryption that cannot be compromised, aiming to protect critical data from online dangers. Furthermore, secure communication channels are being revolutionized through quantum communication systems that employ quantum key distribution principles, crucial for exchanges in diplomacy and defense sectors. In addition, advancements in the fields of quantum sensors and computing bear considerable importance for national safety by bolstering intelligence collection and threat assessment capabilities. As security infrastructures increasingly incorporate these quantum technologies ethical issues such as privacy rights, surveillance practices, and the possibility of militarization must lead policy deliberations to guarantee their use is both transparent and ethically grounded within the international security domain.

Quantum Computing and Cryptography in Defense

Within the defense domain, the convergence between quantum computing and cryptography marks a significant shift necessitating vigilant adaptation and deliberate planning. The remarkable processing capabilities of quantum computing introduce both prospects and challenges for security sectors. Its facility for rapidly executing intricate computations might transform encryption techniques, bolstering data protection measures. Utilizing fundamental aspects of quantum mechanics such as entanglement and superposition, quantum cryptography promises encryption that cannot be broken, potentially shielding critical defense information against cyber incursions. Yet, this progress prompts moral dilemmas, underscoring the need for thoughtful contemplation on matters concerning data privacy and international security paradigms. As countries delve into the repercussions of deploying quantum technologies within their defense strategies, it becomes crucial to develop stringent ethical guidelines and foster global collaboration to guarantee conscientious yet secure adoption in protecting national stakes.

Nonproliferation of Quantum Weapons

In the current geopolitical scenario, inhibiting the spread of quantum weaponry stands as a pivotal issue, demanding worldwide collaboration and strict control measures to avert the intensification of potential disputes. With the progression of quantum technologies, concocting quantum arms presents moral and security quandaries that necessitate thoughtful scrutiny. Crucial are international accords and solid structures to steer the employment and application of quantum arms, ensuring their non-utilization for nefarious objectives or to disrupt global harmony. Through formulating explicit norms and observation systems, thwarting the proliferation of quantum armaments can be efficaciously orchestrated, fostering tranquility and protection in an ever-evolving technological realm. The execution of overtly monitored mandates is crucial in defending against aberrations in using quantum faculties and preserving equilibrium in worldwide safety frameworks. The moral repercussions associated with atomic affray tools beckon unified actions from global entities to diminish hazards and maintain ethical benchmarks within technical evolution spheres.

International Agreements and Regulations

Regulatory frameworks and international accords are pivotal in managing the ethical utilization and responsible deployment of advancing quantum technological fields. The rapid growth in areas such as quantum sensing, cryptography, and computing highlights an urgent requirement for worldwide collaboration to tackle emerging security vulnerabilities and moral dilemmas. Crafting schemes for global partnerships is critical to leveraging quantum advancements favorably for community benefits while neutralizing risks. It falls upon policy architects to untangle the intricacies involved in overseeing these technologies, aiming at bolstering data protection, openness, and digital safeguarding measures. By encouraging a mutual exchange among countries, global contracts can establish norms guiding the innovation's ethical progress along with its application and control within quantum realms—the objective being a framework where novel breakthroughs are synced with consideration toward ethics and collective safety nets.

XL. Quantum Mechanics and Space Exploration

Quantum mechanics, by elucidating the behaviors and foundational aspects of particles, crucially molds our comprehension regarding the endeavor of space exploration. Diving deeper into cosmic realms makes the fundamentals of quantum mechanics ever more pertinent for navigating through intricate systems and occurrences in space. The notions such as superposition and quantum entanglement are on the brink of transforming communication and propulsion technologies within space voyages. This reflection of quantum entities' interconnectedness onto the cosmological scale hints at an overarching unity among celestial structures across the cosmic vastness, accentuating a comprehensive methodology to unravel cosmic secrets. By welcoming both uncertainties and opportunities that quantum mechanics introduces, we stand on the verge of exploring new horizons in outer space adventures, essentially closing the chasm between theoretical intricacies and their practical deployments toward venturing into uncharted territories.

Quantum Sensors in Spacecraft

In spacecraft, the employment of quantum sensors stands as a pioneering utilization of the principles inherent in quantum mechanics, propelling our ventures into the cosmos via detailed recordings and data gathering. By capitalizing on the phenomena of quantum entanglement and superposition, such sensors boast unmatched precision and sensitivity when it comes to monitoring gravitational forces, levels of radiation, and peculiarities in space. These devices herald a significant advancement in space technology, equipping spacecraft with the ability to traverse extensive cosmic expanses with incomparable finesse and dependability. Their knack for perceiving minor variations in their surroundings paves new pathways in astrophysics and cosmology, illuminating aspects previously enshrouded in mystery about our universe. For novices within the domain of quantum mechanics, grasping how these sensors function aboard spacecraft accentuates the palpable influence that theories from this avant-garde discipline have on state-of-the-art technologies; thus cultivating an enriched recognition of its intricacies and capabilities that define this transformative sector.

Quantum Communication in Space

Space-based quantum communication is an advanced implementation of quantum physics, brimming with potential to transform the transmission of information over long distances fundamentally. The distinct characteristics of quantum entanglement could facilitate secure and immediate exchanges across the immense expanses of space, surmounting conventional challenges associated with data security and transmission velocity. Such progress not only exemplifies the tangible repercussions of quantum theories on contemporary devices but also underscores the deep linkages among quantum entities dispersed through space. Employing quantum entanglement for communicating in outer space necessitates exact adjustments and creative methodologies to effectively exploit this extraordinary occurrence. As we further probe into outer space, quantum communication emerges as a pillar of scientific advancement, presenting unmatched prospects for rapid, secure data transfers that might revolutionize our comprehension of both communications technology and the cosmos itself.

Implications for Interstellar Travel

For interstellar voyages in the future, quantum mechanics introduces significant consequences, acting as a portal to transform our grasp of navigating through space. By adopting quantum phenomena such as superposition and entanglement, the possibilities for exceeding light-speed communication and traversing immense cosmic expanses become enticingly achievable. Ideas like quantum teleportation might initiate a new era for starbound expeditions by allowing immediate data transfer and possibly even material itself. The use of quantum entities' linked nature could establish secure, efficient networks for space dialogue, considerably shortening voyage durations and unlocking uncharted territories in distant galaxy exploration. As the enigmas of quantum mechanics unfold, the alluring prospect of traveling among stars shifts from an elusive fantasy towards attainable fact, extending the limits of both our creativity and technical prowess to venture boldly into realms previously unexplored.

XLI. Quantum Mechanics and Philosophy of Science

Delving into the meld of quantum mechanics and science philosophy necessitates a deep reassessment of core principles along with knowledge frameworks. The realities born out of quantum theory upend views that everything is predetermined, sparking deep philosophical musings over existence's essence, consciousness exploration, and the illusion or reality of autonomy. Quantum marvels such as particle-wave dualism, particles' mysterious connections (entanglement), and their ability to exist in multiple states simultaneously reach far past physics' borders, prompting thoughts on universal interconnectivity and how observers effectuate changes in perceived truths. The Copenhagen Interpretation's stance on outcomes being probabilities until observed challenges our grasp on cause-and-effect logic and predestined occurrences. With quantum mechanics progressively redefining scientific exploration boundaries and paving pathways for new technological breakthroughs, discussions shift towards understanding the intricate dance between seeing is believing (scientific realism) versus skepticism about tangible proof (anti-realism). This further highlights the nuanced bond amongst witnessing events unfold coversely interpreting them expansively compared across folds of existence's fabric through observation lenses. Stepping onto this exploratory path aimed at unraveling beginner-level mysteries surrounds in quantum mechanics mandates an integrative approach; fusing theoretical insights application scopes alongside philosophical depths unravels extensive panoramas within this mystifying yet revolutionary domain.

Scientific Realism and Instrumentalism

Contrasting perspectives, such as instrumentalism and scientific realism, delineate the essence of what scientific doctrines and the entities they narrate are thought to represent. Asserting that theories in science strive for a precise depiction of universal truth, scientific realism embraces the notion that entities not directly observable - like particles at the quantum level - possess a real, autonomous being. This outlook is tethered to the conviction that truthful representations of our world emerge through scientific conjecture, thereby enriching our comprehension of nature's mechanisms. Opposingly, instrumentalism advocates for viewing scientific postulates merely as apparatuses aiding in forecasting and elucidating phenomena we can observe; it refrains from asserting these unseen entities' literal existence. For novices untangling quantum mechanics' mysteries, distinguishing between these philosophical attitudes lays down an essential scaffold for decrypting complex quantum behaviors including superposition phenomena, entanglement at a subatomic scale, and dualities between wave and particle characteristics. Delving into how one's perception of quantum notions might be sculpted by either instrumentalism or realism aids fledgling scholars in piercing through quantum mechanics' intricate veneer with philosophically informed lenses.

Theory Change and Scientific Revolutions

Diving into quantum mechanics' universe, we witness pivotal changes in theory and scientific upheavals that redefine our cosmos comprehension. Quantum realities trigger a paradigm overhaul, challenging traditional perspectives and necessitating a rethinking of core tenets. The transition from classical to quantum physics, led by pioneers such as Planck, Bohr, and Heisenberg, marks a monumental shift in scientific ideology. Highlighting the Copenhagen Interpretation's reliance on probability underscores the elusive nature of quantum occurrences prior to observation—contrasting fixed outlooks while promoting an adaptable viewpoint. This upheaval further unfolds through quantum field theory exploration by luminaries like Dirac and Feynman; enlightening us about particle dynamics and natural forces. These progressions do not merely deepen our microscopic realm insights but also reveal significant potential impacts across technology, philosophy, and how we discern reality. Welcoming uncertainty and interconnectedness inherent in quantum mechanics paves the path for deciphering universal secrets and achieving an integrated understanding of being.

Quantum Mechanics and Scientific Method

Delving into the interplay between quantum mechanics and methodology in science, it is clear that the core concepts of quantum theory pose a challenge to established scientific norms. The probabilistic essence and principles of uncertainty inherent in quantum mechanics disrupt the conventional deterministic perspective, prompting a revision in how scientific investigation is conducted. Acknowledging phenomena such as superposition, entanglement, and the dual nature of particles and waves calls for a reassessment of our methods for observing and interpreting events at the microscopic scale. By accepting the complex connections and fluid characteristics of entities within quantum physics, researchers are compelled to deal with the intrinsic vagueness found in these phenomenons, advocating for an adaptable and broad-minded perspective towards conducting research. As investigators strive to understand how quantum mechanics impacts our perception of reality, consciousness, and progress in technology, there emerges a need for adapting the scientific method into one that embraces more complexity and subtlety in comprehension across cosmology.

XLII. Quantum Mechanics and Literature

In the domain of literature, quantum mechanics has echoed profoundly, influencing narratives with its mysterious principles and proposing challenging viewpoints on existence. Authors infuse literary creations with the uncertain, interconnected essence of quantum theories to probe into themes like unpredictability, linkage, and reality's mutable character. Novels such as "Infinite Jest" by David Foster Wallace and Tom Stoppard's theatrical piece "Arcadia" weave in notions from quantum mechanics to investigate intricate aspects of human life alongside the entwined nature of time-space continuums. These narrative ventures do more than amuse; they provoke contemplation among readers regarding quantum mechanics' extensive impact on awareness, autonomy choice-making practices (free will), and existence's core truth. The fusion of science with art through integrating quantum concepts within stories presents a distinctive viewpoint for mulling over universal enigmas and our position relative to them.

Literary Themes Inspired by Quantum Theory

In the literary domain, motifs drawn from quantum theory have intrigued both authors and audiences, serving as a medium to delve into the depths of reality's complexity and existence. By drawing analogies between the unpredictability and connectedness central to quantum mechanics with life's multifaceted aspects, writers create stories that question established views and encourage deep reflection. Themes such as quantum entanglement and superposition resonate within tales through ideas of connection, contrast, and decision-making. Protagonists dealing with varied realities, altering viewpoints, and the fading line between awareness and subconscious states mirror the mystifying aspects of quantum theories. Through subtle narrative techniques and creative examinations, these literary motifs not just amuse but also spark introspection and pondering, echoing quantum theory's significant influence on our grasp of both cosmos' magnitude.

Science Fiction and Quantum Mechanics

The complex ideas of quantum mechanics are deeply entwined with the realm of science fiction, which serves as a speculative canvas for dissecting the enigmas of the quantum zone. By merging imaginative tales with scientific theory, narratives within science fiction probe into aspects of quantum phenomena such as superposition, entanglement, and reality's essence, thereby challenging traditional views and expanding the limits of human comprehension. Fictional works take creative liberties to stretch quantum principles into visions of advanced technologies, alternate dimensions, and possibilities for traversing time, thus engaging viewers and igniting inquisitiveness regarding fundamental scientific doctrines. Through immersion in science fiction mediums, neophytes can cultivate a sophisticated grasp on the intricacies inherent in quantum mechanics within an approachable and relevant framework leading toward a more profound inquiry into essential notions that scaffold our perception of the quantum domain.

Narrative Structures and Quantum Concepts

When delving into the complex narratives interlaced with quantum principles, there emerges a blend of theoretical constructs and philosophical reflections. Quantum mechanics challenges traditional narrative approaches with its intricate concepts like the superposition effect and particle-wave duality, pushing for a reassessment of what we perceive as reality. Narratives have the capacity to intertwine different viewpoints and temporal sequences similarly to how quantum entities coexist in multiple conditions at once, offering an enriched narrative fabric. The weaving together of characters and story arcs reflects the entanglement seen among quantum particles, suggesting that each aspect of a story has the potential to affect others significantly. By adopting elements of uncertainty and changeability within storytelling, it's possible to reflect quantum mechanics' core ideas where possibilities and unpredictabilities dictate the flow of stories. This fusion between narrative techniques and quantum theories prompts a reconsideration of storytelling practices by exploring how deeply intertwined are our notions of reality and fiction.

XLIII. Quantum Mechanics and Gender Studies

Intriguing connections emerge between quantum mechanics and gender studies, putting conventional views on identity, autonomy, and the distribution of power under scrutiny. Delving into quantum notions with a focus on gender allows for an investigation into how concepts like entanglement and superposition can serve as metaphors for the interconnectivity and malleability found in gender identities and presentations. Similar to how entities within quantum physics coexist in numerous states at once, so too might individuals traverse through varied genders simultaneously. The phenomenon of wave-particle duality is reflective of understanding gender as a continuum rather than adhering strictly to a dichotomous framework. Furthermore, the observer's impact within quantum theory encourages a reflection upon how societal standards mold perceptions around gender critically. Integrating perspectives from quantum physics into the discourse on gender elucidates complex processes behind identity development while protesting against fixed views on what defines masculinity or femininity. Such cross-disciplinary methodology not only augments our grasp over esoteric aspects of quantums but also advances discussions surrounding social constructions related to gender roles significantly.

Gender Perspectives in Physics

In the realm of scientific exploration and novelty, including quantum mechanics analysis, gender viewpoints considerably influence the dynamics of research. The domain of physics has historically seen a male majority, causing a noticeable absence in female and minority group representation. It is essential to tackle gender prejudices while fostering diversity within physics to build an environment that is both more inclusive and fair. Injecting gender perspectives into quantum mechanics debates helps in overturning established accounts and elevating varied contributions in the sphere of physics inquiries. Pushing for the inclusion of women and other undervalued groups into physics may unveil novel perceptions and methodologies towards deciphering quantum occurrences. Adopting a stance on gender variety not only amplifies research excellence but also aids in achieving a more rounded and exhaustive comprehension of quantum mechanics alongside its societal repercussions. Committing to advancing gender equity within physics stands to significantly embellish the discipline, setting a foundation for groundbreaking revelations and progressions.

Contributions of Women in Quantum Science

In the domain of quantum physics, the pivotal roles played by women have been instrumental in deepening our grasp on the complex enigmas associated with quantum mechanics. Women have significantly influenced the progression of this field through their initial research efforts and pathbreaking findings. Prominent personalities such as Marietta Blau, who contributed substantially to advancements in particle physics and radioactivity studies, alongside Chien-Shiung Wu, known for her critical experiments that ushered new insights into beta decay and parity violation, stand as testament to women's priceless contributions within quantum science. Their endeavors have broadened our comprehension concerning quantum phenomena whilst simultaneously challenging conventional gender norms prevailing within scientific circles. Embracing female researchers in quantum studies has furnished novel viewpoints, creative methodologies, and an enriched academic environment at large; thus underscoring their indispensable role in unraveling the complexities tied to quantum mechanics for forthcoming cohorts.

Addressing Gender Bias in STEM Fields

For the advancement of a scientific community that's both diverse and inclusive, it is imperative to tackle the issue of gender bias within STEM domains. By acknowledging and confronting entrenched biases and stereotypes, we make way for equal chances across genders. Efforts aimed at mentorship, professional cultivation, and academic initiatives are key to narrowing the gender divide in fields traditionally dominated by males such as engineering and physics. It is crucial to motivate young females towards STEM interests from their formative years while ensuring support for women in both academia and the corporate sphere, driving forward gender equality. Studies have demonstrated that teams with rich diversity tend to be more inventive and achieve greater success, underscoring the necessity of combating gender prejudice for maximizing the scientific sector's potential. Adopting an approach centered on inclusivity within STEM not only aids individuals but also contributes to monumental breakthroughs including innovations pivotal like those found in quantum mechanics – essentially molding science's future trajectory alongside technological advancements.

XLIV. Quantum Mechanics and Social Science

In the domain of social science, the melding of quantum mechanics into its framework offers illuminating revelations about the essence of connectivity and unpredictability in human exchanges and societal frameworks. The phenomenon of quantum entanglement reflects the complex web of interdependent relationships within sociopolitical systems, underscoring a fundamental link among individuals and collectives. Additionally, the Principle of Uncertainty undermines deterministic ideologies that are widespread in social sciences by pointing out the erratic nature inherent in human actions and societal developments. Through harnessing quantum concepts to dissect social occurrences, an enhanced comprehension regarding how decisions are made, how networks function, and how a collective mentality is formed becomes apparent. This shift provides innovative angles on interpersonal dealings and crafting policies. As various disciplines witness transformations due to quantum innovations, their incorporation into social science reveals novel approaches for analyzing and maneuvering through the complicated nexus of individual relations and communal constructs.

Sociological Impact of Quantum Discoveries

The influence of quantum breakthroughs stretches far into sociology, altering perceptions of existence and mutual dependence. Quantum mechanics disrupt conventional ideologies, igniting debates over autonomy, predetermination, and the essence of human awareness. Phenomena such as entanglement and superposition in the quantum realm have direct implications on technological advancements, contributing to the evolution of areas like quantum computing and cryptography. Such progress not only propels industries forward but also raises moral questions about data protection, safety measures, and shifts in worldwide authority structures. The capacity for quantum innovations to reform renewable energy sources heralds significant changes in how we manage waste and achieve energy conservation. Facing the intricacies inherent within quantum mechanics is crucial for leveraging its positive aspects ethically while confronting philosophical challenges and striving towards a balance between scientific discovery and communal principles. Deliberate contemplation followed by active involvement is needed to assimilate quantum leaps with societal standards seamlessly into our day-to-day existence.

Quantum Mechanics in Societal Decision-Making

The enigmatic yet captivating doctrines of quantum mechanics play a crucial role in influencing the methods used for making decisions within society. The concept of interconnectivity and the random nature prevalent in quantum occurrences push against old, predictable viewpoints, pushing towards a more detailed way to tackle intricate matters. By accepting the uncertainties underscored by quantum mechanics, those in charge can employ strategies that are both adaptable and flexible while facing various hurdles. Furthermore, ideas such as superposition and entanglement from quantum theories provide fresh insights into how things are interconnected and reveal numerous potential outcomes, illuminating complex social interactions' nuances. Venturing into quantum mechanics not only broadens our comprehension of cosmic entities but also unveils important understandings regarding human actions and relations, directing us to a mindful method for sculpting our shared destiny. Through meditation on the principles found within quantum fields, leaders could enhance innovation, Promote steadfastness, and nurture understanding—elevating their ability to creatively resolve societal issues with an assorted perspective. These efforts inevitably lead toward groundbreaking solutions rooted firmly within sustainability.

Interdisciplinary Research and Collaboration

Advancing our grasp of quantum mechanics and its applied uses significantly depends on interdisciplinary research and cooperative efforts. The creation of partnerships among physicists, engineers, and mathematicians introduces fresh viewpoints, which catalyze the development of inventive resolutions and pioneering breakthroughs. This cross-disciplinary strategy facilitates the merging of varied skill sets, enhancing the study of quantum characteristics such as entanglement, superposition, and the wave-particle duality. Collaborative ventures enable scholars to address intricate issues in fields like quantum computing, cryptography, and communication frameworks, setting the stage for revolutionary technological advancements. The collaboration across various fields not only hastens scientific achievements but also fosters an environment replete with shared insights and reciprocal admiration—vital for extending the frontiers of quantum investigations and molding the tech-centric future.

XLV. Quantum Mechanics and Ethics

Venturing into the elaborate sphere of Quantum Mechanics, alongside its moral ramifications, unveils a dynamic interrelation between scientific exploration and ethical judgments. Probing into the deep connectedness of quantum entities unsettles conventional views on reality and instigates a reconsideration of moral frameworks. Notions such as quantum entanglement and superposition amplify the intricacy in making decisions and how observations affect quantum systems. The moral facets tied to quantum innovations like computing and cryptography bring up crucial worries over privacy, protection, and international influence balances, urging for an astute moral guideline to steer their creation and utilization thoughtfully. Handling these ethical issues is crucial to confirm that advancements in quantum technology enrich society while adhering strictly to ethical standards. This debate highlights the importance of weaving ethical contemplations within both studying Quantum Mechanics' applications earnestly aiming for an engagement with these revolutionary technologies mindful of their moral undertones.

Ethical Implications of Quantum Technologies

Navigating the ethical terrain created by quantum innovations demands meticulous attention for responsible engagement in areas like quantum computing and cryptography. These technological leaps hold great promise for transforming diverse sectors, but they bring with them concerns over data protection, security risks, and wider societal ramifications that must be thoroughly tackled to guarantee their ethical deployment. The unbreakable encryption potential of quantum cryptography through the distribution of quantum keys calls for deep-seated ethical structures to prevent its exploitation and ensure the privacy of individuals is preserved. As we see an increasing incorporation of quantum technologies into everyday life, it becomes crucial for lawmakers, scholars, and communities at large to initiate dialogues aimed at forging standards that equipoise groundbreaking advancements with moral contemplation. This will steer us towards a future where the fruits of quantum progress are harnessed beneficially across society whilst maintaining high bars for ethically mindful innovation and use.

Responsibility in Scientific Research

In the realm of scientific inquiry, holding to a code of conduct is critical, more so in intricate areas such as quantum mechanics where the paths of ethical questions and technological progress often cross. Those exploring the subtle aspects of quantum phenomena are obligated to maintain honesty, uprightness, and moral principles to guarantee their discoveries are represented accurately and used responsibly. With the rise of quantum-based innovations like computing and cryptography influencing myriad sectors, it becomes even more vital that developers possess an ethical vision within their work's framework. The challenge lies in aligning knowledge exploration with understanding its potential effects on society; this requires a judicious and deliberate stance towards both discovery and application in science. Committing to moral scrutiny while fostering transparent discussions about how quantum mechanics affects our world empowers researchers to tread through this complex domain with morality and anticipation at the forefront—thereby laying down a foundation for a future where science not only advances but does so thoughtfully with regard to its broader impact.

Ethical Education for Quantum Scientists

In the realm of quantum scientist preparation, an essential element is ethical tutoring, pivotal for maneuvering through the intricacies and possible moral predicaments that emerge in the frontier of innovative exploration. Ethical principles lay down a cornerstone for conscientious scientific investigation, steering investigators towards upholding rectitude, openness, and answerability in their endeavors. Through weaving ethical tutelage into quantum scientists' education framework, academies may imbed a profound consciousness and obligation regarding ethics among emerging researchers. This might cover dialogues concerning moral consequences tied to quantum tech advancements, issues around safeguarding data privacy and security, along with morally utilizing quantum progress across different sectors. In essence, educating on ethics furnishes quantum experts with the know-how to engage in enlightened decision-making that cherishes moral norms while fostering the accountable development of quantum innovations amid a swiftly changing backdrop.

XLVI. Quantum Mechanics and Language

Through exploring the captivating crossroad of Language and Quantum Mechanics, a deep connection surfaces that shines light on the fundamental nature of how we communicate. The words we select to depict and make sense of quantum phenomena's complexities act as an essential link connecting abstract theoretical ideas with concrete comprehension. As quantum mechanics upsets traditional perceptions of existence, so too must our linguistic constructs be reshaped to contain the counterintuitive actions of quantum objects. The detailed lexicon, exact terminologies, and symbolic representations in quantum mechanics are critical in forming our understanding of its unseen universe. Language transforms into a vehicle for articulating complex concepts such as superposition, entanglement, and the dual aspects of particles and waves to novices, decoding the enigmas surrounding quantum theory into coherent revelations. By harnessing language effectively as a mechanism for sharing knowledge about quantum notions, we clear a path towards enriching insight and admiration for this mysterious domain; crossing linguistic obstacles unfolds the wonders held within quantum mechanics for those eager to learn.

Terminology and Conceptual Understanding

Plunging into the sphere of quantum mechanics at an introductory level necessitates a foundational comprehension of certain terminologies and conceptual knowledge to effectively maneuver through its intricacies. Understanding terminology acts as a vital tool for unlocking the deep-seated concepts that form the basis of quantum mechanics, facilitating individuals in grasping the core principles dictating particle behavior at the quantum scale. Acquainting oneself with jargon such as superposition, entanglement in quantum contexts, and duality between waves and particles allows novices to start deciphering this intriguing domain's secrets. These ideas act as cornerstone elements for further inquiry into the enigmatic aspects of quantum existence, illuminating phenomena that cast doubt on conventional perceptions of reality. It is essential for beginners to develop a robust conceptual grasp on these terms to understand fully quantum mechanics' nature and its repercussions on science's wider vista. With straightforward elucidations and captivating illustrations, newcomers are positioned to embark upon a voyage filled with discovery, setting up opportunities for advanced understanding of intricate principles ruling over the realm governed by quantum laws.

Language as a Tool for Teaching Quantum Physics

As a potent mechanism, language unravels the complex notions of quantum physics to novices venturing into comprehending this intricate domain. By deploying straightforward and lucid language, teachers can efficiently transmit elusive concepts such as superposition, entanglement in quantum mechanics, duality of wave-particle nature, and the wave equation in an intelligible fashion. Employing relatable instances along with visual tools like imaginative experiments and schematic representations enables pupils to internalize these essential tenets more successfully. This method not only improves comprehension but also cultivates an enriched admiration for the revolutionary hypotheses proposed by pioneers like Max Planck and Niels Bohr. Leveraging linguistic capabilities as an educational device, instructors can diminish the chasm dividing quantum theories from beginner scholars, thus facilitating a broader and more captivating delve into the quantum dimension.

Communication of Quantum Ideas to the Public

The conveyance of quantum theories to the populace is vital in unraveling intricate scientific notions and promoting engagement with state-of-the-art breakthroughs. In making fundamental principles of quantum mechanics understandable for novices, being clear and approachable is essential to narrow the divide between scholarly research and general comprehension. By simplifying core ideas such as superposition, entanglement in quantum physics, and the duality of waves and particles through easy-to-relate examples and visual aids, people can comprehend basic premises of quantum philosophy. Fostering an environment that welcomes open discussion and inventive uses of multimedia can elevate understanding and pique interest in the phenomena related to quantum among varied groups. Through joint endeavors amongst educators, physicists, and the lay audience, a broader yet insightful conversation regarding quantum mechanics is possible; this enhances overall scientific acumen while enabling individuals to delve into the marvels of Quantum Universe with assurance.

XLVII. Quantum Mechanics and Psychology

The fusion of Quantum Mechanics with Psychology unlocks deep observations into consciousness and human actions, breaking through the conventional barriers of scientific exploration. The unpredictable elements and interconnectedness introduced by quantum phenomena refute the notion of determinism, aligning with psychology's examination of free will and awareness. In quantum mechanics, the observer effect reflects how human perception is inherently subjective, underscoring how observation alters reality - a concept that echoes within psychological studies on perception and thought processes. By welcoming the uncertainties and convolution inherent in quantum mechanics, psychologists can enhance their grasp on mental functions and behaviors, forging pathways for novel research endeavors and theoretical progressions. This merging heralds an enticing prospect to investigate how quantum existence intertwines with psychological dynamics, illuminating the complex ties between the minuscule world of quantum entities and the vast intricacies of human awareness and activities.

Cognitive Approaches to Quantum Concepts

Exploring Quantum Concepts through Cognitive Lenses provides an intriguing peek into how quantum mechanics and cognitive processes intertwine, highlighting the interaction between human mental constructs and the intricate aspects of quantum events. By investigating both cognitive psychology and theories of quantum mechanics, scholars seek to narrow the divide that exists between our natural comprehension of the vast macro world and the seemingly paradoxical behaviors observed in quantum phenomena. Through research focusing on how we make decisions, recognize patterns, and construct mental frameworks, these cognitive methodologies reveal our perception and interpretation of complex quantum notions such as entanglement and superposition. Such strategies do not only improve our understanding of quantum physics but also lay down foundational blocks for creative instructional techniques aimed at novices and those outside specialized fields, promoting a more profound acknowledgment and grasp of fundamental concepts essential to quantum physics among a broader demographic. The interplay between cognitive studies and quantum science offers potential solutions for unraveling the mysteries surrounding quantal concepts thereby rendering them more comprehensible to avid learners; this serves to nurture a community that is better informed about—and thus keener to explore—the mystifying domain of quantum mechanics.

Psychological Impact of Quantum Discoveries

The mind's reaction to quantum breakthroughs goes beyond simple scientific inquiry, probing deep into the core of human awareness and how we perceive what's real. Revealing quantum mechanics with its complex ideas such as superposition and entanglement of particles, defies old beliefs and prompts people to reevaluate existence itself. These insights compel a shift from fixed outcomes thinking towards accepting uncertainty as intrinsic to cosmos fabric. As novices navigate through the dense territory of quantum theory, they encounter a unity among all entities and realize monumental effects on human thought process. The mental unease caused by reflecting on concepts like the dual nature of waves and particles, along with how observation influences reality, ignites self-reflection and philosophical deliberations over autonomy, fate versus free will, and the complexities woven into our universe's tapestry. Acknowledging the mental influence of uncovering quantum phenomena can transform an individual's comprehension about cosmology and oneself significantly; it cultivates an enriched respect for enigmas that transcend traditional understanding.

Quantum Mechanics in Psychological Theory

The implications of Quantum Mechanics for Psychological Theory are fascinating, upsetting the apple cart of classical deterministic perspectives on human cognition and behavior. The woven-together aspect of quantum bits, illustrated by marvels such as entanglement and superposition, hints at a novel framework for grasping consciousness. Ideas concerning the observer effect and how observation molds reality ignite debates over how human consciousness could sway quantum systems. This crossroad between quantum mechanics and psychological theory stirs musings on autonomy versus fate, along with the essence of what's real, providing an innovative angle on how our thoughts and choices might be swayed by quantum events. Probing into what quantum mechanics means for psychology could reveal fresh understandings about the nexus between mind and body plus the intricacies inherent in human actions. This exploration aims to lay groundwork for groundbreaking methods in comprehending and addressing mental health conditions.

XLVIII. Quantum Mechanics and the Arts

'The Intersection of Quantum Mechanics with Artistic Creativity' unveils an absorbing blend between the domains of science and imaginative prowess, highlighting the significant influence that quantum notions wield on artistic evolution and perception. As it ventures into understanding reality's complex essence, quantum mechanics is utilized by creators to examine themes such as mutual dependence, unpredictability, and indeterminacy within their productions. Observing through wave-particle duality perspectives, artists make vague the distinction between sensory experience and physicality, urging spectators to revise their grasp on shape and purpose. The concept of quantum entanglement heralds tales of unified destinies absent geographic closeness in literary works and visual artistry-inciting narratives about knitted existences across distances. The inherent volatility and adaptability found in quantum dimensions urge a reconsideration of accepted artistic practices—this encourages critical reflections concerning existence's authentic nature alongside consciousness within creative expressions. By intertwining explorative scientific endeavors with artful manifestations, 'Quantum Mechanics with Artistic Creativity' fosters enriched reverence towards universal enigmas along with omnipresent connectedness—eroding disciplinary divides while provoking deep meditation and marveling admiration.

Intersections Between Quantum Physics and Artistic Expression

In the intriguing overlap of quantum mechanics and artistic creativity, we observe a space where quantum physics' elusive ideas meet artists' interpretative endeavors. The capability of art to encapsulate visually the intricacies and connections found in quantum occurrences is unparalleled. Quantum entanglement, superposition, and the duality of waves-particles serve as muses for artists aiming to produce works that redefine conventional views and prompt deep reflections on reality's essence. Employing novel methods and materials, these creators navigate the ambiguousness and malleability inherent in quantum states, softening the divide between scientific inquiry and artistic practice. By embedding principles from quantum theory into their artworks, such individuals not only translate complex scientific notions for wider appreciation but also incite discussions about being's core truths and sentient awareness. This confluence between artistry influenced by quantum physics carves out an enticing domain for delving into universal enigmas via imaginative expression—encouraging self-examination while broadening perceptions regarding both science's factual territory and art's expressive expanse.

Art as a Medium for Explaining Quantum Ideas

Utilizing art as an enthralling method, complex quantum notions become clearer to a wider audience, narrowing the division between everyday comprehension and scientific theories. Through artistic visuals, the elusive concepts of superposition and entanglement in quantum mechanics are transformed into more graspable and intriguing subjects for novices. Artistic rendition of wave-particle duality through vibrant and imaginative expressions enhances understanding while igniting curiosity about how quantum elements are intertwined. This inventive strategy aids in unraveling daunting scientific axioms while encouraging pondering over the deep-seated effects of quantum mechanics within our cosmos. Employing artistry to elucidate quantum ideas not only simplifies education but also amplifies acknowledgment of the intricacy and elegance inherent in the quantum realm, thus proving itself as a crucial pedagogic aid for newcomers delving into the mysteries of quantum mechanics.

Collaborations Between Artists and Physicists

The cooperation of physicists with artists unveils an intriguing blend where the realm of scientific precision meets the world of creative imagination, possibly creating a conduit from the complex theories underpinning quantum mechanics to the palpable sphere of artistry. Through such partnerships, complex ideas from science can be morphed into visual forms that capture wider attention and ignite inquisitiveness. By picturing phenomena such as superposition or entanglement within artistic frameworks, these joint efforts provide insights that extend beyond conventional explanations found in science, offering fresh vistas for understanding and narrative building. Such mutual associations delve deeper into how nature interweaves, opening new paths for both imaginative exploration and nuanced scientific discourse. Artists inject a novel vibrancy into convoluted scientific notions, aiding in unraveling the mysteries surrounding quantum characteristics to make them comprehensible even for novices—thereby cultivating a more comprehensive grasp on quantum dimensions.

XLIX. Quantum Mechanics and Innovation

The fusion of innovation and quantum mechanics unveils a groundbreaking capacity for enhancing contemporary scientific insights and technologies. Quantum mechanics upends traditional perspectives on reality with notions such as entanglement and superposition, igniting revolutions in sectors like encryption and computing. Further muddling our comprehension of the cosmos are the Uncertainty Principle along with the concept of wave-particle duality, which compels a reassessment of our views on reality's fabric and our connectedness within it. Transitioning to quantum field theory from quantum mechanics enriches our understanding of the minute realm, shedding light on nature's forces and particle interactions. As advancements in quantum cryptography and computing continue to extend limitations, these developments beckon us to weather uncertainties while leveraging quantum phenomena's fluidity for both cutting-edge innovations and ethical deliberations alike.

Driving Technological Advances

Navigating the frontier of technological progression with quantum mechanics as a guide opens up a universe of capability that transforms computation, communication, and encoding methodologies. Phenomena such as superposition and entanglement in quantum realms defy conventional ideologies, introducing unparalleled velocity and defense mechanisms in the realm of data manipulation and dissemination. Such theoretical insights find practical resonance across areas like medicine and ecological studies, where quantum detectors improve both accuracy and productivity. Although it embeds intricacies, quantum mechanics equally acts as a muse for creativity, stretching the imaginable horizons across diverse sectors. Grasping these foundational principles of quantum theory aids not merely in pioneering technical progress but also enriches our introspective musings over existence, awareness, and mutual dependence. By acknowledging indeterminacy and harnessing the vast possibilities offered by quantum innovations, transformative strides are made towards establishing an era dominated by quantum capabilities that surpass present constraints.

Quantum Innovation Ecosystems

Ecosystems dedicated to quantum innovation are crucial for the progression of quantum technologies and their applications, molding the contours of contemporary science and technology. These synergistic configurations include concerted interplay between universities, research entities, governmental bodies, and corporate collaborators, propelling strides in areas such as quantum computing, communication, and encryption. Encouraging an interdisciplinary collaborative mode of operation enables these ecosystems to cultivate a milieu where ideas flow freely among participants while sharing resources and savoir-faire essential for addressing intricate obstacles within quantum research and evolution. Drawing upon the intellectual prowess of luminaries like Max Planck, Niels Bohr, and Werner Heisenberg helps thrust forward seminal discoveries and advances within the realm of quantum mechanics. Whether it's delving into the mysteries surrounding quantum entanglement or employing superposition principles in quantum computation efforts—such ecosystems spearhead development at science's cutting-edge as well as technological progressions leading towards a transformative era marked by substantial shifts across diverse sectors thanks to potentialities inherent within a future dominated by quantum phenomena.

Fostering Creativity in Quantum Research

Encouraging innovative thought in the realm of quantum studies is essential for advancing the frontier of scientific exploration and technological progress. By fostering an attitude that accepts uncertainty and delves into non-traditional concepts, scientists can unveil new understandings and possible innovations within quantum physics. This strategy facilitates probing into fresh ideas and theoretical constructs that might contest well-established models, culminating in revolutionary progressions in domains such as quantum informatics and secure communication. The essence of creativity within quantum investigations entails looking past conventional limits, acknowledging the complex interrelations inherent to quantum events, and employing creative strategies for addressing challenges. Cultivating an atmosphere of inventiveness and receptiveness enables investigators to exploit the vast capabilities of quantum physics fully, laying the groundwork for groundbreaking enhancements that will define the trajectory of scientific and technological advancement.

L. Quantum Mechanics and the Future

At the vanguard of scientific investigation, quantum mechanics flags the dawn of a novel epoch marked by technological advancements and deep philosophical introspection. By unlocking the complexities surrounding quantum states, entanglement, and the dual nature of particles and waves, we forge new avenues toward an era where enhancements in computation, modes of communication, and perhaps even our understanding of consciousness await. Transitioning from classical to quantum physics through the efforts led by luminaries such as Planck and Bohr has established a groundwork for quantum-based innovations that are reshaping various sectors. The potential held by quantum computing to deliver vastly superior processing capabilities alongside the unbreakable security offered by quantum cryptography exemplifies its limitless applications. Yet, the journey into the essence of quantum mechanics further incites us to explore unknown dimensions pertaining to quantum consciousness and how observation intertwines with reality itself. Venturing through this landscape filled with ambiguity and interconnectedness expands our view on what is achievable in both science and societal constructs alike, propelling us towards confronting enigmatic queries uncovered by exploring quantum mechanics.

Emerging Technologies and Their Impact

In the burgeoning field of quantum mechanics, novel technologies are transforming our comprehension of the cosmos and leading innovations across multiple sectors. The doctrines of superposition and entanglement in quantum mechanics have been instrumental in propelling significant breakthroughs in areas like quantum computing and encryption, which offer unmatched levels of security and computational capabilities. These advancements are not just revolutionizing industry landscapes but also altering our conceptualization of existence and the cosmic interrelation. Delving into the theoretical underpinnings of quantum mechanics reveals complex phenomena such as particle-wave duality and the impact observation has on outcomes, immersing us in the enigmatic realms of the quantum domain. Grasping these emerging technologies along with their deep-seated influence on both scientific inquiry and societal development demands an appreciation for ambiguity and dynamism in how we perceive reality, thus setting a foundation for a future rich with infinite potentialities anchored in quantum innovation.

Quantum Mechanics in Future Societies

Future societies stand on the brink of a transformation, guided by quantum mechanics, which promises to overhaul technological environments and question established norms. Delving into this complex domain brings to light critical notions such as superposition, entanglement in quantum realms, and the dual nature of particles and waves. These central ideas not only deepen our comprehension of the cosmos but also pave the way for revolutionary progress in areas like quantum computing and secure cryptographic methods. This heralds an epoch marked by enhanced security in communication protocols and data handling procedures. Quantum mechanics' applications transcend its theoretical underpinnings, offering practical resolutions to current dilemmas; it suggests a future dominated by the mingled existence and indefiniteness of quantum elements propelling our tech advances and introspective philosophical debates. Adopting the intricacies that come with understanding quantum mechanics opens up boundless opportunities for altering societal structures dramatically, cementing its role as a pivotal element in forthcoming scientific research and innovation projects.

Speculations on the Evolution of Quantum Science

Conjectures regarding the development of quantum science unfold a mosaic of interwoven breakthroughs that have transformed contemporary physics and technology. By exploring the historical origins of quantum mechanics, seminal figures such as Max Planck, Niels Bohr, and Werner Heisenberg established the foundations for a paradigm transformation in how we comprehend the cosmos. Ideas like quantum states, entanglement, and the Principle of Uncertainty have not only contested conventional beliefs but also advanced progress in computing and encryption strategies, setting the stage for innovations in quantum-based technologies like cryptography. As quantum mechanics continues to explore its boundaries with complex yet captivating phenomena such as particle-wave duality and entanglement at a quantum level, this speculative voyage into its progression incites reflections on what lies ahead in terms of scientific inquiry and technological breakthroughs. The mystical attributes of quantum science summon us to consider its revolutionary capabilities, encouraging novices to dive into an absorbing universe filled with potentialities and revelations.

LI. Conclusion

To sum up, when we probe the complex dimensions of quantum mechanics, it becomes clear that the workings of the universe adhere to a set of principles starkly different from what our everyday experiences might lead us to believe. Notions such as superposition, entanglement in quantum physics, and the duality between waves and particles defy our conventional grasp on reality while paving new paths for investigation in both physics and technological innovation. The historical evolution of quantum theory through the efforts of notable scientists including Planck, Bohr, and Heisenberg has laid down groundwork for transformative breakthroughs in domains like computing and cryptographic technologies. The inherent uncertainties and intricate nature of quantum mechanics compel us to welcome vagueness and reconsider our predilection for deterministic interpretations, advocating for an adaptative yet exploratory stance towards scientific endeavors. Through understanding the interrelated aspects of quantum entities along with recognizing significant impacts brought upon by quantum phenomena across philosophical discourse and technological advancements, we find ourselves equipped to tread into the ambiguities bestowed by the realm of quantum science with eager curiosity coupled with refreshed comprehension.

Summary of Key Points Discussed

As beginners venture into the complexities of quantum mechanics, it's clear that grasping its core concepts is crucial for understanding this complicated area. The exploration of quantum phenomena challenges traditional perspectives on reality with elements such as entanglement in quantum physics, the duality of waves and particles, and Heisenberg's Uncertainty Principle. Icons like Planck, Bohr, and Heisenberg illuminated these key theories, transforming physics and technological development by establishing vital foundations for progress in areas like quantum computing and secure communication through cryptography. Highlighted by the Copenhagen Interpretation is the inherent randomness in outcomes within quantum systems, stimulating debates regarding fate versus voluntariness. Delving into these depths reveals a web of interconnected facts that question longstanding beliefs while laying fertile ground for novel breakthroughs promising thrilling possibilities for science-based futures and tech advancements.

Reflection on the Role of Quantum Mechanics in Science and Technology

Deliberation on quantum mechanics' influence within science and technology outlines the critical transformation owing to quantum foundations in contemporary novelties. Initiated by pioneers such as Max Planck and Niels Bohr, quantum mechanics has ushered a revolution across diverse sectors via notions like superposition, entanglement, and the dual aspect of waves and particles. These core tenets have facilitated developments in areas of quantum computing, cryptography, and systems for communication, overturning conventional perceptions of existence while expanding technological limits. In delving into the intertwined characteristics of quantum objects coupled with an acceptance of indeterminacy within scientific exploration, quantum mechanics not only amplifies tech applications but also incites philosophical ponderings about causal relations, autonomy in choice making alongside universe's essence comprehension. Incorporating quantum mechanics into science plus technology heralds a shift in paradigm that nurtures an intensified grasp over reality along with complexity inherent to cosmos leading innovation moreover carving prospective paths.

Future Prospects and Directions in Quantum Research

The outlook and pathways forward in the study of quantum phenomena are brimming with potential to redefine diverse sectors while advancing the frontiers of scholarly inquiry. With progress in areas such as quantum computation and cryptographic methods, there lies a horizon ripe for transformation in how we process data, communicate, and maintain privacy. The foundational principles of quantum mechanics not only contest conventional perceptions of existence but also present actionable strategies for addressing intricate calculative challenges. As investigations into quantum realms press onwards, the collaborative spirit spanning physics, engineering, and computer science paves roads toward conquering novel technological territories and sparking innovation. Acknowledging the inherent uncertainties and interconnectedness that mark quantum occurrences will prove critical in steering forthcoming scholarly pursuits while maximizing the vast capabilities that quantum theory offers in redesigning tomorrow's scientific vista. This melding of theoretical constructs with tangible implementations signals an era poised on converting what once seemed inconceivable into reality, thereby heralding groundbreaking strides in research and technological breakthroughs.

Bibliography

Aleksandr Akovlevich Khinchin. 'Mathematical Foundations of Information Theory.' Courier Corporation, 1/1/1957

Nick Jones. 'The Observer Effect.' Blackstone Publishing, 3/15/2022

Mohammad Reza Pahlavani. 'Selected Topics in Applications of Quantum Mechanics.' BoD – Books on Demand, 5/13/2015

Division of Behavioral and Social Sciences and Education. 'Science Teaching Reconsidered.' A Handbook, National Research Council, National Academies Press, 3/12/1997

Hans Radder. 'The Material Realization of Science.' From Habermas to Experimentation and Referential Realism, Springer Science & Business Media, 5/3/2012

Norma T. Mertz. 'Theoretical Frameworks in Qualitative Research.' Vincent A. Anfara, Jr., SAGE Publications, 10/30/2014

Peter van Loock. 'Quantum Teleportation and Entanglement.' A Hybrid Approach to Optical Quantum Information Processing, Akira Furusawa, John Wiley & Sons, 5/3/2011

Stewart Wilson. 'All About.' Secure Communication, CreateSpace Independent Publishing Platform, 2/14/2016

Saverio Pascazio. 'Quantum Communication and Quantum Networking.' First International Conference, QuantumComm 2009, Naples, Italy, October 26-30, 2009, Revised Selected Papers, Alexander Sergienko, Springer, 1/8/2010

Giulio Casati. 'Principles Of Quantum Computation And Information: A Comprehensive Textbook.' Giuliano Benenti, World Scientific, 12/12/2018

Federico Grasselli. 'Quantum Cryptography.' From Key Distribution to Conference Key Agreement, Springer Nature, 1/4/2021

National Intelligence Council. 'Global Trends 2040.' A More Contested World, COSIMO REPORTS, 3/1/2021

Division on Engineering and Physical Sciences. 'Quantum Computing.' Progress and Prospects, National Academies of Sciences, Engineering, and Medicine, National Academies Press, 4/27/2019

Wolfgang H. Polak. 'Quantum Computing.' A Gentle Introduction, Eleanor G. Rieffel, MIT Press, 8/29/2014

Henry Kyambalesa. 'Marketing in the 21st Century: Concepts, Challenges and Imperatives.' Concepts, Challenges and Imperatives, Routledge, 11/1/2017

Jon B. Hagen. 'Radio-Frequency Electronics.' Circuits and Applications, Cambridge University Press, 6/11/2009

Mario Pérez-Montoro. 'The Phenomenon of Information.' A Conceptual Approach to Information Flow, Scarecrow Press, 6/11/2007

Mohsen Razavy. 'Quantum Theory of Tunneling.' World Scientific, 1/1/2003

Division of Behavioral and Social Sciences and Education. 'Learning and Understanding.' Improving Advanced Study of Mathematics and Science in U.S. High Schools, National Research Council, National Academies Press, 8/6/2002

R. Shankar. 'Principles of Quantum Mechanics.' Springer Science & Business Media, 12/6/2012

James C. Robinson. 'An Introduction to Ordinary Differential Equations.' Cambridge University Press, 1/8/2004

Paul Urban. 'The Schrödinger Equation.' Proceedings of the International Symposium "50 Years Schrödinger Equation" in Vienna, 10th–12th June 1976, Walter Thirring, Springer Science & Business Media, 12/6/2012

Robert W. Seabloom. 'Engineering Fundamentals.' In Measurements, Probability, Statistics, and Dimensions, Keith C. Crandall, McGraw-Hill, 1/1/1970

Jeff Sanny. 'University Physics Volume 3.' Samuel J. Ling, Samurai Media Limited, 12/19/2017

Carlo Rovelli. 'Helgoland.' Making Sense of the Quantum Revolution, Penguin, 5/24/2022

Shan Gao. 'The Meaning of the Wave Function.' In Search of the Ontology of Quantum Mechanics, Cambridge University Press, 3/16/2017

Alfred C Ewing. 'The Fundamental Questions of Philosophy (Routledge Revivals).' Routledge, 4/3/2013

LeeAnn Racz. 'Handbook of Measurements.' Benchmarks for Systems Accuracy and Precision, Adedeji B. Badiru, CRC Press, 10/8/2018

W. Heisenberg. 'Nuclear Physics.' Open Road Media, 5/7/2019

David Lindley. 'Uncertainty.' Knopf Doubleday Publishing Group, 2/12/2008

Abdellatif Zaidi. 'Information Theory for Data Communications and Processing.' Shlomo Shamai (Shitz), MDPI, 1/13/2021

F. Selleri. 'The Einstein, Podolsky, and Rosen Paradox in Atomic, Nuclear, and Particle Physics.' Alexander Afriat, Springer Science & Business Media, 11/11/2013

Ralph Barton Perry. 'General Theory of Value.' Its Meaning and Basic Principles Construed in Terms of Interest, Longmans, Green, 1/1/1926

Jed Brody. 'Quantum Entanglement.' MIT Press, 2/18/2020

Letitia Meynell. 'Thought Experiments in Philosophy, Science, and the Arts.' Mélanie Frappier, Routledge, 1/1/2013

Gabriel Popescu. 'Principles of Biophotonics.' Linear systems and the Fourier transform in optics. Volume 1, IOP Publishing, 1/1/2018

Karol Życzkowski. 'Geometry of Quantum States.' An Introduction to Quantum Entanglement, Ingemar Bengtsson, Cambridge University Press, 8/18/2017

Joshua Isaacson. 'Quantum Computing for the Quantum Curious.' Ciaran Hughes, Springer Nature, 3/22/2021

Michael de Podesta. 'Understanding the Properties of Matter.' CRC Press, 5/18/2020

Oswaal Editorial Board. 'Oswaal NTA NEET (UG) PLUS Supplement for Additional Topics(Physics, Chemistry, Biology) and 10 Mock Test Papers, Updated As Per New Syllabus (Set of 2 Books) For 2024 Exam.' Oswaal Books, 12/5/2023

G. Alwyn Zittrauer. 'An Epic of Metaphysical Existence.' BOT Publishing, 1/1/2005

Franco Selleri. 'Wave-Particle Duality.' Springer Science & Business Media, 12/6/2012

Robert B. Sanders. 'Contributions of African American Scientists to the Fields of Science, Medicine, and Inventions.' Nova Publishers, Incorporated, 1/1/2015

Helge Kragh. 'Quantum Generations.' A History of Physics in the Twentieth Century, Princeton University Press, 3/24/2002

Jeff Suzuki. 'Mathematics in Historical Context.' MAA, 8/27/2009

Lucile Vaughan Payne. 'The Lively Art of Writing.' Words, Sentences, Style and Technique -- an Essential Guide to One of Today's Most Necessary Skills, Penguin, 3/1/1969

Karl A. Van Bibber. 'Edward Teller Centennial Symposium.' Modern Physics and the Scientific Legacy of Edward Teller : Livermore, CA, USA, 28 May 2008, Stephen B. Libby, World Scientific, 1/1/2010

R. Mirman. 'Quantum Mechanics, Quantum Field Theory.' Geometry, Language, Logic, iUniverse, 12/1/2004

David J. Griffiths. 'Introduction to Quantum Mechanics.' Cambridge University Press, 1/1/2017

www.ingramcontent.com/pod-product-compliance
Lightning Source LLC
Chambersburg PA
CBHW050206230526
45470CB00001B/263